Praise for *F*

T0200794

"None make the case for the final divorce of religion and science, with permanent restraining orders against harassment and stalking of science by religion, better than Coyne."　　　　—Ray Olson, *Booklist* (starred review)

"An important book that deserves an open-minded readership."
　　　　　　　　　　　　　　　　　　　　　　　　—*Kirkus Reviews*

"Many people are confused about science—about what it is, how it is practiced, and why it is the most powerful method for understanding ourselves and the universe that our species has ever devised. In *Faith Versus Fact*, Coyne has written a wonderful primer on what it means to think scientifically, showing that the honest doubts of science are better—and more noble—than the false certainties of religion. This is a profound and lovely book. It should be required reading at every college on earth."
　　　　　　—Sam Harris, author of *The End of Faith*, *The Moral Landscape*, and *Waking Up*

"The distinguished geneticist Jerry Coyne trains his formidable intellectual firepower on religious faith, and it's hard to see how any reasonable person can resist the conclusions of his superbly argued book. Though religion will live on in the minds of the unlettered, in educated circles faith is entering its death throes. Symptomatic of its terminal desperation are the 'apophatic' pretensions of 'sophisticated theologians,' for whose empty obscurantism Coyne reserves his most devastating sallies. Read this book and recommend it to two friends."　　　　—Richard Dawkins, author of *The God Delusion*

"The truth is not always halfway between two extremes: some propositions are flat wrong. In this timely and important book, Jerry Coyne expertly exposes the incoherence of the increasingly popular belief that you can have it both ways: that God (or something God-ish, God-like, or God-oid) sort-of exists; that miracles kind-of happen; and that the truthiness of dogma is somewhat-a-little-bit-more-or-less-who's-to-say-it-isn't like the truths of science and reason."
　　　　—Steven Pinker, Johnstone Family Professor of Psychology, Harvard University; author of *The Better Angels of Our Nature*

ABOUT THE AUTHOR

Jerry A. Coyne is Professor Emeritus in the Department of Ecology and Evolution at the University of Chicago, where he specialized in evolutionary genetics. His *New York Times* bestseller, *Why Evolution Is True*, was one of *Newsweek*'s "50 Books for Our Times" in 2010.

Faith
VERSUS
FACT

WHY SCIENCE AND RELIGION
ARE INCOMPATIBLE

Jerry A. Coyne

PENGUIN BOOKS

PENGUIN BOOKS

An imprint of Penguin Random House LLC
375 Hudson Street
New York, New York 10014
penguin.com

First published in the United States of America by Viking Penguin,
an imprint of Penguin Publishing Group, a division
of Penguin Random House LLC, 2015
Published in Penguin Books 2016

THE LIBRARY OF CONGRESS HAS CATALOGED THE
HARDCOVER EDITION AS FOLLOWS:

Coyne, Jerry A., 1949–
Faith versus fact : why science and religion are incompatible / Jerry A. Coyne.
pages cm
Includes bibliographical references and index.
ISBN 9780670026531 (hc.)
ISBN 9780143108269 (pbk.)
1. Religion and science.   2. Faith and reason.   I. Title.
BL240.3.C69 2015
201'.65—dc23
2015001103

Printed in the United States of America
5   7   9   10   8   6

Set in Warnock Pro
Designed by Amy Hill

To Bruce Grant, my first mentor in science,
and
Małgorzata, Andrzej, and Hili Koraszewscy, for providing
a warm, secular haven for thinking and writing

God is an hypothesis, and, as such, stands in need of proof: the onus probandi [burden of proof] rests on the theist.

—*Percy Bysshe Shelley*

We have already compared the benefits of theology and science. When the theologian governed the world, it was covered with huts and hovels for the many, palaces and cathedrals for the few. To nearly all the children of men, reading and writing were unknown arts. The poor were clad in rags and skins—they devoured crusts, and gnawed bones. The day of Science dawned, and the luxuries of a century ago are the necessities of to-day. Men in the middle ranks of life have more of the conveniences and elegancies than the princes and kings of the theological times. But above and over all this, is the development of mind. There is more of value in the brain of an average man of to-day—of a master-mechanic, of a chemist, of a naturalist, of an inventor, than there was in the brain of the world four hundred years ago.

These blessings did not fall from the skies. These benefits did not drop from the outstretched hands of priests. They were not found in cathedrals or behind altars—neither were they searched for with holy candles. They were not discovered by the closed eyes of prayer, nor did they come in answer to superstitious supplication. They are the children of freedom, the gifts of reason, observation and experience—and for them all, man is indebted to man.

—*Robert Green Ingersoll*

CONTENTS

The Genesis of This Book

*The good thing about science is that it's true
whether or not you believe in it.*

—Neil deGrasse Tyson

In February 2013, I debated a young Lutheran theologian on a hot-button topic: "Are science and religion compatible?" The site was the historic Circular Congregational Church in Charleston, South Carolina, one of the oldest churches in the American South. After both of us gave our twenty-minute spiels (she argued "yes," while I said "no"), we were asked to sum up our views in a single sentence. I can't remember my own précis, but I clearly recall the theologian's words: "We must always remember that faith is a gift."

This was one of those *l'esprit d'escalier,* or "wit of the staircase," moments, when you come up with the perfect response—but only well after the opportunity has passed. For shortly after the debate was over, I not only remembered that *Gift* is the German word for "poison," but saw clearly that the theologian's parting words undercut her very thesis that science and religion are compatible. Whatever I actually said, what I *should* have said was this: "Faith may be a gift in religion, but in science it's poison, for faith is no way to find truth."

This book gives me a chance to say that now. It is about the different ways that science and religion regard faith, ways that make them incompatible for discovering what's true about our universe. My thesis is that religion and science compete in many ways to describe reality—they both make "existence claims" about what is real—but use different tools to meet this goal. And I argue that the toolkit of science, based on reason and empirical study, is reliable, while that of religion—including faith, dogma, and revelation—is unreliable and leads to incorrect, untestable, or conflicting conclusions. Indeed, by relying on faith rather than evidence, religion renders itself *incapable* of finding truth.

I maintain, then—and here I diverge from the many "accommodationists" who see religion and science, if not harmonious or complementary, at least as not in conflict—that religion and science are engaged in a kind of war: a war for understanding, a war about whether we should have good reasons for what we accept as true.

Although this book deals with the conflict between religion and science, I see this as only one battle in a wider war—a war between rationality and superstition. Religion is but a single brand of superstition (others include beliefs in astrology, paranormal phenomena, homeopathy, and spiritual healing), but it is the most widespread and harmful form of superstition. And science is but one form of rationality (philosophy and mathematics are others), but it is a highly developed form, and the only one capable of describing and understanding reality. All superstitions that *purport* to give truths are actually forms of pseudoscience, and all use similar tactics to immunize themselves against disproof. As we'll see, advocates of pseudosciences like homeopathy or ESP often support their beliefs using the same arguments employed by theologians to defend their faith.

While the science-versus-religion debate is one battle in the war between rationality and irrationality, I concentrate on it for several reasons. First, the controversy has become more widespread and visible, most likely because of a new element in the criticism of religion. The most novel aspect of "New Atheism"—the form of disbelief that distinguishes the views of writers like Sam Harris and Richard Dawkins from the "old" atheism of people like Jean-Paul Sartre and Bertrand Russell—is the observation that most religions are grounded in claims that can be regarded as *scientific*. That is, God, and the tenets of many religions, are *hypotheses* that can, at least in principle, be examined by science and reason. If religious claims can't be substantiated with reliable evidence, the argument goes, they should, like dubious scientific claims, be rejected until more data arrive. This argument is buttressed by new developments in science, in areas like cosmology, neurobiology, and evolutionary biology. Discoveries in those fields have undermined religious claims that phenomena like the origin of the universe and the existence of human morality and consciousness defy scientific explanation and are therefore evidence for God. Seeing their bailiwick shrinking, the faithful have become more insistent that religion is actually a way of understanding nature that complements science. But the most important reason to concentrate on religion rather than other forms of irrationality is not to document a historical conflict, but because, among all forms of superstition, religion has by far the most potential for public harm. Few are damaged by belief in astrology; but, as we'll see in the final chapter, many have been harmed by belief in a particular god or by the idea that faith is a virtue.

I have both a personal and a professional interest in this argument, for I've spent my adult life teaching and studying evolutionary biology, the brand of science most vilified and rejected by religion. And a bit more biography is in order: I was raised as a secular Jew, an upbringing that, as most people know, is but a hairsbreadth from atheism. But my vague beliefs in a God were abandoned almost instantly when, at seventeen, I was listening to the Beatles' *Sergeant Pepper* album and suddenly realized that there was simply no evidence for the religious claims I had been taught—or for anybody else's, either. From the beginning, then, my unbelief rested on an absence of evidence for anything divine. Compared with that of many believers, my rejection of God was brief and painless. But after that I didn't think much about religion until I became a professional scientist.

There's no surer route to immersion in the conflict between science and religion than becoming an evolutionary biologist. Nearly half of Americans reject evolution completely, espousing a biblical literalism in which every living species, or at least our own, was suddenly created from nothing less than ten thousand years ago by a divine being. And most of the rest believe that God *guided* evolution one way or another—a position that flatly rejects the naturalistic view accepted by evolutionary biologists: that evolution, like all phenomena in the universe, is a consequence of the laws of physics, without supernatural involvement. In fact, only about one in five Americans accepts evolution in the purely naturalistic way scientists see it.

When I taught my first course in evolution at the University of Maryland, I could hear the opposition directly, for in the plaza right below my classroom a preacher would often hold forth loudly about how evolution was a tool of Satan. And many of my own students, while dutifully learning about evolution, made it clear that they didn't believe a word of it. Curious about how such opposition could exist despite the copious evidence for evolution, I began reading about creationism. It was immediately evident that virtually all opposition to evolution comes from religion. In fact, among the dozens of prominent creationists I've encountered, I've known of only one—the philosopher David Berlinski—whose view isn't motivated by religion.

Finally, after twenty-five years of teaching, facing pushback all the way, I decided to address the problem of creationism in the only way I knew: by writing a popular book laying out the evidence for evolution. And there were mountains of evidence, drawn from the fossil record, embryology, molecular biology, the geography of plants and animals, the development and construction of animal bodies, and so on. Curiously, nobody had written such a book. Practical people, I figured—or even skeptical ones—would surely come around to accepting the scientific view of evolution once they'd seen the evidence laid out in black and white.

I was wrong. Although my book, *Why Evolution Is True*, did well (even nosing briefly onto the *New York Times* bestseller list), and although I received quite a few letters from religious readers telling me I'd "converted" them to evolution, the proportion of creationists in America didn't budge: for thirty-two years it's hovered between 40 and 46 percent.

It didn't take long to realize the futility of using evidence to sell evolution to Americans, for faith led them to discount and reject the facts right before their noses. In my earlier book I recounted the "aha" moment when I realized this. A group of businessmen in a ritzy suburb of Chicago, wanting to learn some science as a respite from shoptalk, invited me to talk to them about evolution at their weekly luncheon. I gave them a lavishly illustrated lecture about the evidence for evolution, complete with photos of transitional fossils, vestigial organs, and developmental anomalies like the vanishing leg buds of embryonic dolphins. They seemed to appreciate my efforts. But after the talk, one of the attendees approached me, shook my hand, and said, "Dr. Coyne, I found your evidence for evolution very convincing—but I still don't believe it."

I was flabbergasted. How could it be that someone found evidence convincing but *was still not convinced*? The answer, of course, was that his religion had immunized him against my evidence.

As a scientist brought up without much religious indoctrination, I couldn't understand how *anything* could blinker people against hard data and strong evidence. Why couldn't people be religious and still accept evolution? That question led me to the extensive literature on the relationship between science and religion, and the discovery that much of it is indeed what I call "accommodationist": seeing the two areas as compatible, mutually supportive, or at least not in conflict. But as I dug deeper, and began to read theology as well, I realized that there were intractable incompatibilities between science and religion, ones glossed over or avoided in the accommodationist literature.

Further, I began to see that theology turns itself, or at least the truth claims religion makes about the universe, into a kind of science, but a science using weak evidence to make strong statements about what is true. As a scientist, I saw deep parallels between theology's empirical and reason-based justifications for belief and the kind of tactics used by pseudoscientists to defend their turf. One of these is an a priori commitment to defend and justify one's preferred claims, something that stands in strong contrast to science's practice of constantly testing whether its claims might be wrong. Yet religious people were staking their very lives and futures on evidence that wouldn't come close to, say, the kind of data the U.S. government requires before approving a new drug for depression. In the end I saw that the claims for the compatibility of science and religion were weak, resting on assertions about the nature of religion that few believers really accept, and that religion could never be made compatible with science without diluting it so seriously that it was no longer religion but a humanist philosophy.

And so I learned what other opponents of creationism could have told me: that persuading Americans to accept the truth of evolution involved not just an education in facts, but a de-education in faith—the form of belief that replaces the need for evidence with simple emotional commitment. I will try to convince you that religion, as practiced by most believers, is severely at odds with science, and that this conflict is damaging to science itself, to how the public conceives of science, and to what the public thinks science can and cannot not tell us. I'll also argue that the claim that religion and science are complementary "ways of knowing" gives unwarranted credibility to faith, a credibility that, at its extremes, is responsible for many human deaths and might ultimately contribute to the demise of our own species and much other life on Earth.

Science and religion, then, are competitors in the business of finding out what is true about our universe. In this goal religion has failed miserably, for its tools for discerning "truth" are useless. These areas are incompatible in precisely the same way, and in the same sense, that rationality is incompatible with irrationality.

Let me hasten, though, to add a few caveats.

First, some "religions," like Jainism and the more meditation-oriented versions of Buddhism, make few or no claims about what exists in the universe. (I'll shortly give a definition of "religion" so that my thesis becomes clear.) Adherents to other faiths, like Quakers and Unitarian Universalists, are heterogeneous, with some "believers" being indistinguishable from agnostics or atheists who practice a nebulous but godless spirituality. As the beliefs of such people are often not *theistic* (that is, they don't involve a deity that interacts with the world), there is less chance that they will conflict with science. This book deals largely with theistic faiths. They're not the totality of religions, but they constitute by far the largest number of religions—and believers—on Earth.

For several reasons I concentrate on the Abrahamic faiths: Islam, Christianity, and Judaism. Those are the religions I know most about, and, more important, are the ones—particularly Christianity—most concerned with reconciling their beliefs with science. While I discuss other faiths in passing, it is mostly the various brands of Christianity that occupy this book. Likewise, I will talk mostly about science and religion in the United States, for here is where their conflict is most visible. The problem is less pressing in Europe because the proportion of theists, particularly in northern Europe, is much lower than in America. In the Middle East, on the other hand, where Islam is truly and deeply in conflict with science, such discussions are often seen as heretical.

Finally, there are some versions of even the Abrahamic religions whose tenets are so vague that it's simply unclear whether they conflict with science. Apophatic, or "negative," theology, for instance, is reluctant to make claims about the nature or even the existence of a god. Some liberal Christians speak of God as a "ground of being" rather than as an entity with humanlike feelings and properties that behaves in specified ways. While some theologians claim that these are the "strongest" notions of God, they have that status only because they make the fewest claims and are thus the least susceptible to refutation—or even discussion. For anyone having the least familiarity with religion, it goes without saying that such watered-down versions of faith are not held by most people, who accept instead a personal god who intervenes in the world.

This brings us to the common claim that critics of religion accept a "straw man" fallacy, seeing all believers as fundamentalists or scriptural literalists, and that we neglect the "strong and sophisticated" versions of faith held by liberal theologians. A true discussion of faith/science compatibility, this argument runs, demands that we deal only with these sophisticated forms of belief. For if we construe "religion" as simply "the beliefs of the average believer," then arguing that those beliefs are incompatible with science is just as nonsensical as construing "science" as the rudimentary and often incorrect understanding of science held by the average citizen.

But this parallel is wrong in several ways. First, while many laypeople hold erroneous views of science, they neither *practice* science nor are considered part of the scientific community. In contrast, the average believer not only practices religion but may also belong to a religious community that may try to spread its beliefs to the wider society. Further, while theologians may know more about the *history* of religion—or the work of other theologians—than do regular believers, they have no special expertise in discerning the nature of God, what he wants, or how he interacts with the world. In understanding the claims of their faith, "regular" religious believers are far closer to theologians than are science-friendly laypeople to the physicists and biologists they admire. Throughout this book I'll consider the claims both of garden-variety believers and of theologians, for while the problem of faith versus science is most serious for the regular believer, it is the theologians who use academic arguments to convince believers that their faith is compatible with science.

I emphasize that my claim that science and religion are incompatible does not mean that most religious people reject science. Even evolution, the science most scorned by believers, is accepted by many Jews, Buddhists, Christians, and liberal Muslims. And, of course, most believers have no problem with the idea of supernovas, photosynthesis, or gravity. The conflict plays out in only a few specific areas of science, but also in the validation of faith in general. My argument for incompatibility deals not with people's *perceptions*, but with the contradictory ways that science and religion support their claims about reality.

I begin by showing evidence that the conflict between religion and science is substantial and widespread. This evidence includes the incessant production of books and official statements by both scientists and theologians assuring us that there really is compatibility, by using different and sometimes contradictory arguments. The sheer number and diversity of these assurances suggest that there's a problem that hasn't been resolved. Further evidence for conflict includes the high proportion of scientists in both the United States and the United Kingdom who are atheists, a proportion of nonbelievers roughly ten times higher than that in the general public. Also, in America and other countries, there are laws that privilege faith by giving it precedence over science, as in the medical treatment of one's children. Finally, the existence of pervasive creationism, as well as widespread belief in religious and spiritual healing, shows an obvious conflict between science and religion—or between science and faith.

The second chapter lays out the terms of engagement: the ways I construe science and religion, and what I mean by "incompatibility." I'll argue that the incompatibility operates at three levels: methodology, philosophy, and outcomes—what "truths" are uncovered by science versus faith.

Chapter 3 takes on accommodationism, analyzing a sample of the arguments used by both religious people and scientific organizations to argue for a harmony between science and faith. The two most common arguments are the existence of religious scientists, and Stephen Jay Gould's prominent idea of "non-overlapping magisteria" (NOMA), in which science encompasses the domain of facts about the universe while religion occupies the orthogonal realm of meaning, morals, and values. In the end, all accommodationist strategies fail because they don't resolve the huge disparity between discerning "truths" using reason versus faith. I'll describe three examples of the problems that arise when scientific advances flatly contradict religious dogma: theistic (God-guided) evolution, claims about the existence of Adam and Eve, and Mormon beliefs about the origin of Native Americans.

The fourth chapter, "Faith Strikes Back," tackles not only the ways that religion is said to *contribute* to science, but also the way the faithful *denigrate* science as a way of defending their own turf. The arguments are diverse, and include claims that science actually supports the idea of God by supplying answers to questions supposedly beyond the ken of science. I call these endeavors the "new natural theology"—a modern version of eighteenth- and nineteenth-century arguments that purported to show the hand of God in nature. The updated arguments deal with the purported "fine-tuning" of the universe—the claimed improbability that the laws of physics would permit the appearance of life—as well as with the claimed inevitability of human evolution, and the details of human morality that, it's argued, resist scientific but not religious explanations. I also take up the notion of "other ways of knowing": the contention that science isn't the only way of ferreting out nature's truths. I'll argue that in fact science is the *only* way to find such truths—if you construe "science" broadly. Finally, I deal with believers' *tu quoque* accusations that science is either derived from religion or afflicted with the same problems as religion. These accusations are also diverse: science is actually a product of Christianity; science involves untestable assumptions, and is therefore based on faith; science is fallible; science promotes "scientism," the view that nonscientific questions are uninteresting; and—the ultimate redoubt of believers—the assertion that while religion has sometimes been harmful, so has science, which has given us things like eugenics and nuclear weapons.

Why should we care whether science and religion are compatible? The last chapter answers this question, showing why reliance on faith, when reason and evidence are available, has created immense harms, including many deaths. The clearest examples involve religiously based healing, which, protected by American law, has killed many, including children who have no choice in their treatment. Likewise, opposition to stem cell research and vaccination, as well as denial of global warming, is sometimes based on religious grounds. I argue that in a world where people must support their opinions with evidence and reason rather than faith, we would experience less conflict over issues like assisted suicide, gay rights, birth control, and sexual morality. Finally, I discuss whether it's *ever* useful to have faith. Are there times when it's all right to hold strong beliefs that are supported by little or no evidence? Even if we can't prove the claims of faith, isn't religion useful as a form of social glue and a wellspring of public morality? Is it possible for science and religion to have a constructive dialogue about these things?

I am aware that criticizing religion is a touchy endeavor (a classic dinner-table no-no), invoking strong reactions even from those who aren't believers but see faith as a societal good. Beyond summarizing what this book is, then, I should also explain what it is not.

Although I deal largely with religion, my purpose is not to show that religion has, on balance, been a malign influence on society. While I do believe this, and in the last chapter emphasize some of the problems of faith, it would be foolish to deny that religion has motivated many acts of goodness and charity. It has also been a solace for the inevitable sorrows of human life, and an impetus for helping others. In the end, it's impossible to perform the "good versus bad" calculus of religion by integrating over history.

My main thesis is narrower and, I think, more defensible: understanding reality, in the sense of being able to use what we know to predict what we don't, is best achieved using the tools of science, and is never achieved using the methods of faith. That is attested by the acknowledged success of science in telling us about everything from the smallest bits of matter to the origin of the universe itself—compared with the abject failure of religion to tell us anything about gods, including whether they exist. While scientific investigations converge on solutions, religious investigations diverge, producing innumerable sects with conflicting and irresolvable claims. Using the predictions of science, we can now land space probes not only on distant planets, but also on distant comets. We can produce "designer drugs" to target a specific individual's cancer, decide which flu vaccines are most likely to be effective in the coming season, and figure out how to finally wipe scourges like smallpox and polio from our planet. Religion, in contrast, can't even tell us if there's an afterlife, much less anything about its nature.

The true harm of accommodationism is the weakening of our organs of reason by promoting useless methods of finding truth, especially that of faith. As Sam Harris notes:

> The point is not that we atheists can prove religion to be the cause of more harm than good (though I think this can be argued, and the balance seems to me to be swinging further toward harm each day). The point is that religion remains the only mode of discourse that encourages grown men and women to pretend to know things they manifestly do not (and cannot) know. If ever there were an attitude at odds with science, this is it. And the faithful are encouraged to keep shouldering this unwieldy burden of falsehood and self-deception by everyone they meet—by their coreligionists, of course, and by people of differing faith, and now, with startling frequency, by scientists who claim to have *no* faith.

In arguing that science is the only way we can really learn things about our universe, I am not calling for a society completely dominated by science, which most people see as a robotic world lacking emotion, empty of art and literature, and devoid of the human need to feel part of something larger than oneself—a need that draws many to religion. Such a world would indeed be sterile and joyless. Rather, I'd claim that adopting a more broadly scientific viewpoint not only helps us make better decisions, both for ourselves and for society as a whole, but also brings alive the many wonders of science barred to those who see it as something distant and forbidding (it's not). What could be more entrancing than understanding at last where we (and all other species) came from, a subject that I've studied all my life? Most important, there would be no devaluating of the emotional needs of humans. I live my life according to the principles I recommend in this book, but if you met me at a party you'd never guess I was a scientist. I am at least as emotional, and enamored of the arts, as the next person, am easily brought to tears by a good movie or book, and do my best to help the less fortunate. All I lack is faith. One can meet all the emotional requisites of a human—except for the assurance that you'll find a life after death—without the superstitions of religion.

Nevertheless, I won't discuss how to replace religion when—as I believe will inevitably happen—it largely disappears from our world. Solutions inevitably depend on the emotional needs of individual personalities, and those interested in such solutions should consult Philip Kitcher's excellent book *Life After Faith: The Case for Secular Humanism*.

Finally, I don't discuss the historical, evolutionary, and psychological origins of religion. There are dozens of hypotheses for how religious belief got started and why it persists. Some invoke direct evolutionary adaptations, others by-products of evolved features like our tendency to attribute events to conscious agents, and still others the usefulness of faith as a societal glue or a way to control others. Definitive answers aren't obvious, and in fact may never be forthcoming. To explore the many secular theories of religion, one should begin with Pascal Boyer's *Religion Explained* and Daniel Dennett's *Breaking the Spell*.

I will have achieved my aim if, by the end of this book, you demand that people produce good reasons for what they believe—not only in religion, but in any area in which evidence can be brought to bear. I'll have achieved my aim when people devote as much effort to choosing a system of belief as they do to choosing their doctor. I'll have achieved my aim if the public stops awarding special authority about the universe and the human condition to preachers, imams, and clerics simply because they are religious figures. And above all, I'll have achieved my aim if, when you hear someone described as a "person of faith," you see it as criticism rather than praise.

Faith Versus Fact

CHAPTER 1

The Problem

For we often talked of my daughter, who died of the fever at fall.
And I thought 'twere the will of the Lord, but Miss Annie she
said it was drains.
                                        —Alfred, Lord Tennyson

T here are no heated discussions about reconciling sport and religion, literature and religion, or business and religion; the important issue in today's world is the harmony between *science* and religion. But why, of all human endeavors that we could compare with religion, are we so concerned with its harmony with science?

The answer, to me at least, seems obvious. Science and religion—unlike, say, business and religion—are competitors at discovering truths about nature. And science is the only field that has the ability to disprove the truth claims of religion, and has done so repeatedly (the creation stories of Genesis and other faiths, the Noachian flood, and the fictitious Exodus of the Jews from Egypt come to mind). Religion, on the other hand, has no ability to overturn the truths found by science. It is this competition, and the ability of science to erode the hegemony of faith—but not vice versa—that has produced the copious discussion of how the two areas relate to each other, and how to find harmony between them.

One can in fact argue that science and religion have been at odds ever since science began to exist as a formal discipline in sixteenth-century Europe. Scientific advances, of course, began well before that—in ancient Greece, China, India, and the Middle East—but could conflict with religion in a public way only when religion assumed both the power and the dogma to control society. That had to wait until the rise of Christianity and Islam, and then until science produced results that called their claims into question.

And so in the last five hundred years there have been conflicts between science and faith—not continuous conflict, but occasional and famous moments of public hostility. The two most notable ones are Galileo's squabble with the church and his sentence to lifetime house arrest in 1632 over his claim of a Sun-centered solar system, and the 1925 Scopes "Monkey Trial" involving a titanic clash between Clarence Darrow and William Jennings Bryan over whether a Tennessee high-school teacher could tell his students that humans had evolved (the jury ruled no). Although both of these incidents have been recast by accommodationist theologians and historians as not involving genuine conflict between science and religion—it's always construed as "politics," "power," or "personal animosity"—the religious roots of these disputes are clear. But even setting these episodes aside, there are many times when churches decried or even slowed scientific advances, episodes recounted in the two books I'll describe shortly. (Of course, churches sometimes *promoted* scientific advances as well: during the advent of smallpox vaccination, churches were on both sides of the issue, with some arguing that it was a social good, others that it was short-circuiting God's power over life and death.)

But these episodes of conflict didn't give rise to public discussion about the relationship of science and religion. That had to wait until the nineteenth century, and was probably ignited by Charles Darwin's 1859 publication of *On the Origin of Species*. The greatest scripture-killer ever penned, the book demolished (not deliberately) an entire series of biblical claims by demonstrating that purely naturalistic processes—evolution and natural selection—could explain patterns in nature previously explainable only by invoking a Great Designer.

And so the modern discussion that science and religion are at odds, with science having the stronger weapons, began with two books published in the late nineteenth century. Historians of science see them as having launched the "conflict thesis": the idea that religion and science are not only at war, but have been *perpetually* at war, with religious authorities opposing or suppressing science at every turn, and science struggling to free itself from the grip of faith. After recounting what they saw as historical clashes between the church and scientists, the authors of both books declared science the victor.

The pugnacity of these works, unusual for their time, was fully expressed in the first: *History of the Conflict Between Religion and Science* (1875) by the American polymath John William Draper:

> Then has it in truth come to this, that Roman Christianity and Science are recognized by their respective adherents as being absolutely incompatible; they cannot exist together; one must yield to the other; mankind must make its choice—it cannot have both.

As the quote implies, Draper saw Catholicism, rather than religion as a whole, as the main enemy of science. This was because of that religion's predominance, the elaborate nature of its dogma, and its attempt to enforce that dogma by civil power. Further, in the late nineteenth century, anti-Catholicism was a dominant strain among the American gentry.

A History of the Warfare of Science with Theology in Christendom, published in 1896, was longer, more scholarly, and more complex in both origin and intent. Its author, Andrew Dickson White, was another polymath—a historian, a diplomat, and an educator. He was also the first president of Cornell University in Ithaca, New York. When White and his benefactor, Ezra Cornell, organized the university in 1865, the state bill describing its mission required that the board of trustees not be dominated by members of any one religious sect, and that "persons of every religious denomination, or of no religious denomination, shall be equally eligible to all offices and appointments." Such secularism was almost unique for that era.

White, a believer, argued that this plurality was actually intended to promote Christianity: "So far from wishing to injure Christianity, we [he and Cornell, who was a Quaker] both hoped to promote it; but we saw in the sectarian character of American colleges and universities, as a whole, a reason for the poverty of the advanced instruction then given in so many of them." This was an explicit attempt to set up an American university on the European model, fostering free inquiry by eliminating religious dogma.

This plan backfired. The secular intent of White and Cornell angered many believers, who accused White of pushing Darwinism and atheism and promoting a curriculum too heavy on science. And they even allowed atheists on the faculty! (Some observers felt that *every* professor should be a pastor.) White's attempt to try "sweet reasonableness" failed, and ultimately he came to view his struggle for university secularism—which he won—as one battle in a wider war between science and theology:

> Then it was that there was borne in upon me a sense of the real difficulty— the antagonism between the theological and scientific view of the universe and of education in relation to it.

This led to thirty years of research culminating in his two-volume work, which was thorough (going far beyond the researches of his predecessor Draper), divisive, and a bestseller. It remains in print today. Despite its catalog of religious opposition to linguistic research, biblical scholarship, medical issues like vaccination and anesthesia, improvements in public health, evolution, and even lightning rods, White insisted that his aim was not to show conflict between science and religion, but only between science and "dogmatic theology." In the end, he hoped—in vain—that his book would actually strengthen religion by calling out its unwarranted incursions into social and natural sciences. In this way it foreshadowed Stephen Jay Gould's accommodationist arguments for the "non-overlapping magisteria" of science and religion, a thesis we'll encounter later.

What White's and Draper's books *did* accomplish was to provide a nucleus for discussing the conflict between science and faith, which in turn raised the ire of theologians and historians of science, who proceeded to argue that the "conflict thesis" was simply wrong. Some historians of science claimed that White's and Draper's scholarship was poor (yes, they did make some errors and omit some countervailing observations, but not nearly enough to invalidate the books' theses), and also that a true reading of the relationship between religion and science showed that they often were in harmony. The rejections of Darwin's and Galileo's theories were, said these historians, exceptions in a genial history of church-science relations, and at any rate those skirmishes were motivated not by religion but by politics or personal quarrels. Indeed, many scientific advances were said to be *promoted* by religious belief, and science itself was touted as a product of the Christianity that permeated medieval Europe.

The truth lies between Draper and White on one hand and their critics on the other. While it's undeniable that religion was important in opposing some scientific advances like the theory of evolution and the use of anesthesia, others, like smallpox vaccination, were both opposed and promoted on biblical grounds. On the other hand, it's a self-serving distortion to say that religion was not an important issue in the persecutions of Galileo and John Scopes. Nevertheless, because not all religions are opposed to science, and much science is accepted by believers, the view that science and faith are perpetually locked in battle is untrue. If that's how one sees the "conflict thesis," then that hypothesis is wrong.

But my view is not that religion and science have always been implacable enemies, with the former always hindering the latter. Instead, I see them as making overlapping claims, each arguing that it can identify truths about the universe. As I'll show in the next chapter, the incompatibility rests on differences in the methodology and philosophy used in determining those truths, and in the outcomes of their searches. In their eagerness to debunk the claims of Draper and White, their critics missed the underlying theme of both books: the failure of religion to find truth about *anything*—be it gods themselves or more worldly matters like the causes of disease.

So what is the evidence that not all is well on the science-and-religion front? For one thing, if the two areas have been found compatible, discussion about their harmony should have ended long ago. But in fact it's growing.

Let's start with a few telling statistics. WorldCat, founded in 1971, is the world's largest compilation of published items, cataloging more than two billion of them in more than seventy thousand libraries worldwide. If you trawl that catalog for books published in English on "science and religion," you'll find a steady increase over the last forty years, from 514 in the decade ending in 1983, to 2,574 in the decade ending in 2013. This doesn't simply reflect the total number of books published, as we can see by normalizing this number by the total number of published books whose subject was "religion." If you do that, the proportion of books on religion that also deal with science has jumped from about 1.1 percent in the former decade to 2.3 percent in the latter. While the number of books on religion nearly doubled between the two decades, the number of books on science and religion increased fourfold. And while not all of the "science and religion" books deal with their relationship, these data support the impression that interest in the topic is growing.

Along with the growth of publications comes a growth in academic courses and programs dealing with science and religion. As Edward J. Larson and Larry Witham noted in 1997, "By one report, U.S. higher education now boasts 1,000 courses for credit on science and faith, whereas a student in the sixties would have long dug in hardscrabble to find even one." Think tanks and academic institutes entirely devoted to science and religion have sprouted; these include the Faraday Institute for Science and Religion at Cambridge University (founded 2006), the Ian Ramsey Center for Science and Religion at Oxford University (founded 1985), and the Center for Theology and the Natural Sciences, in Berkeley, California, founded in 1982 and now boasting of "building bridges between science and theology for 30 years." New academic journals dealing with science and religion have also burgeoned (like *Science, Religion and Culture*, founded in 2014), and, as we'll see below, established scientific organizations have begun to incorporate programs dealing with religion, as well as to issue statements assuring the public that their activities don't conflict with faith.

To a scientist, the clearest sign of disharmony *is* the existence of such programs and statements—for their goal is to try to convince the public that although science and religion might *appear* to be in conflict, they're really not. Why do scientists try to do this? One reason is simply what I call the "nice guy syndrome": a lot more people will like you if you say good things about religion than if you are critical of it. Asserting that your science doesn't step on religion's toes is one way to stay in the good graces of the American public, and everyone else's.

Further, there are those who simply don't like conflict—the "people of good will," as the late paleontologist Stephen Jay Gould called them. For this group, accommodationism seems a reasonable way to avoid conflict, like prohibiting talk about religion and politics at the dinner table. Harmonizing religion and science makes you seem like an open-minded and reasonable person, while asserting their incompatibility makes enemies and brands you as "militant." The reason is clear: religion occupies a privileged place in our society. Attacking it is off-limits, although going after other supernatural or paranormal beliefs like ESP, homeopathy, or political worldviews is not. Accommodationism is not meant to defend science, which can stand on its own, but to show that in some way religion can still make credible claims about the world.

But the real reasons why scientists promote accommodationism are more self-serving. To a large extent, American scientists depend for their support on the American public, which is largely religious, and on the U.S. Congress, which is equally religious. (It's a given that it's nearly impossible for an open atheist to be elected to Congress, and at election time candidates vie with one another to parade their religious belief.) Most researchers are supported by federal grants from agencies like the National Science Foundation and the National Institutes of Health, whose budgets are set annually by Congress. To a working scientist, such grants are a lifeline, for research is expensive, and if you don't do it you could lose tenure, promotions, or raises. Any claim that science is somehow in conflict with religion might lead to cuts in the science budget, or so scientists believe, thus endangering their professional welfare.

These concerns affect all scientists, but evolutionary biologists have an extra worry. Many of our allies in the battle against creationism are liberal religious believers who themselves proclaim that evolution doesn't violate their faith. In court cases brought against public schools that teach creationism, there is no witness more convincing than a believer who will testify that evolution is consonant with his own religion *and* that creationism is not science. Were scientists to say what many of us feel—that religious belief is truly at odds with science—we would alienate these allies and, as many warn us, impede the acceptance of evolution by a public already dubious about Darwin. But there's no hard evidence for either this view or the claim that scientists endanger their livelihood by criticizing faith.

Nevertheless, steeped in a religious culture, many scientific associations prefer to play it safe, proclaiming that science can coexist happily with religion. One example is the American Association for the Advancement of Science's Dialogue on Science, Ethics, and Religion program, devoted to "[facilitating] communication between scientific and religious communities." The "communication" promoted by this largest of America's scientific organizations is always positive; there are no dialogues pointing out any conflicts between science and faith. Likewise, the World Science Festival, a yearly multimedia expo in New York City, always includes a panel or lecture on the compatibility of science and religion. Francis Collins, once head of the Human Genome Project and now director of the National Institutes of Health—and a born-again evangelical Christian—founded BioLogos, an organization devoted to helping antievolution evangelicals retain their faith in Jesus while accepting evolution at the same time. Unfortunately, its success has been limited. It's no coincidence that all three of these programs were funded by grants from the John Templeton Foundation, a wealthy organization founded by a mutual fund billionaire whose dream was to show that science could give evidence for God. As we'll learn shortly, the Templeton Foundation and its huge financial resources are the impetus for many programs promoting accommodationism.

Like BioLogos, the Clergy Letter Project aims to convince believers that evolution does not violate their faith. In this case, religious leaders and theologians have written letters and manifestos affirming that evolution is not heretical. The National Center for Science Education, the nation's most important organization for fighting the spread of creationism, has a "Science and Religion" program with aims identical to those of the Clergy Letter Project. But all of this activity raises a question: if evolution comports so easily with religion, why do we need incessant public proclamations of harmony?

Yet the proclamations keep coming. Here are two. The first is from the American Association for the Advancement of Science:

> The sponsors of many of these state and local proposals [to limit or eliminate the teaching of evolution in public schools] seem to believe that evolution and religion conflict. This is unfortunate. They need not be incompatible. Science and religion ask fundamentally different questions about the world. Many religious leaders have affirmed that they see no conflict between evolution and religion. We and the overwhelming majority of scientists share this view.

Note that this statement, although issued by a group of scientists, is essentially about theology, implying that "true" religions need not conflict with science. But because many Americans believe otherwise—including the 42 percent of the populace that accepts young-Earth creationism—this is in effect telling nearly half the American public that they misunderstand their faith. Groups of scientists clearly have no business declaring what is and is not a "proper" religion.

Here's a declaration from the National Center for Science Education:

> The science of evolution does not make claims about God's existence or non-existence, any more than do other scientific theories such as gravitation, atomic structure, or plate tectonics. Just like gravity, the theory of evolution is compatible with theism, atheism, and agnosticism. Can someone accept evolution as the most compelling explanation for biological diversity, and also accept the idea that God works through evolution? Many religious people do.

But many—perhaps most—religious people *don't*. After all, nearly half of Americans agree with the statement that "God created human beings pretty much in their present form at one time within the last ten thousand years or so." Because nearly 20 percent of Americans are either agnostics or atheists, or say their religion is "nothing in particular," it's a good bet that most religious Americans reject the notion of evolution even in a form guided by God.

The irony in the above statements is that a substantial fraction of *scientists,* and a large majority of accomplished ones, are atheists. Although they have rejected God themselves, presumably because supernatural beings conflict with their evidence-based worldview, many do see religious belief as a social good, but one they don't need themselves. In moments of candor, some scientists admit that these accommodationist statements are really motivated by the personal and political issues I mentioned above.

Similar statements issue from the other side of the aisle. The *Catechism of the Catholic Church,* for instance, states that it's impossible for faith to conflict with fact because both human reason and human faith are vouchsafed by God:

> Though faith is above reason, there can never be any real discrepancy between faith and reason. Since the same God who reveals mysteries and

infuses faith has bestowed the light of reason on the human mind, God
cannot deny himself, nor can truth ever contradict truth. Consequently,
methodical research in all branches of knowledge, provided it is carried
out in a truly scientific manner and does not override moral laws, can
never conflict with the faith, because the things of the world and the
things of faith derive from the same God.

Note the privileging of faith above reason, a bizarre statement that ex-
emplifies the very conflict the church denies. If the two systems must align,
what reason would there be to put one above the other? Further, as we'll see,
the Catholic Church is by and large friendly to evolution, yet many American Catholics are young-Earth creationists, explicitly rejecting the church's
view. What else is that but a discrepancy between faith and reason?

The priority of faith over reason isn't just Catholic policy: it's the view of
many adhering to other religions. A statistic that would frighten any scientist came from a poll of Americans taken in 2006 by *Time* magazine and the
Roper Center. When asked what they would do if science showed that one of
their religious beliefs was wrong, nearly two-thirds of the respondents—64
percent—said that they'd reject the findings of science in favor of their faith.
Only 23 percent would consider changing their belief. Because the pollsters
didn't specify exactly *which* religious belief would conflict with science, this
suggests that the potential conflict between science and religion is not limited to evolution, but could in principle involve any scientific finding that
conflicts with faith. (A prominent one, which we'll discuss later, is the series
of recent scientific discoveries disproving the claim that Adam and Eve
were the two ancestors of all humanity.) A related poll also underscored the
secondary role of scientific evidence for believers: among Americans who
rejected the fact of evolution, the main reasons involved religious belief, not
lack of evidence.

These figures alone cast doubt on statements from religious and scientific organizations that science and religion are compatible. If nearly two-
thirds of Americans will accept a scientific fact only if it's not in clear conflict
with their faith, then their worldview is not fully open to the advances of
science.

Indeed, polls of Americans belonging to various religions, or no religion, show that the *perception* of a conflict between science and faith is widespread. A 2009 Pew poll showed, for instance, that 55 percent of the U.S. public answered "yes" to the question "Are science and religion often in conflict?" (Tellingly, only 36 percent thought that science was at odds with *their own* religious belief.) And, as expected, the perception of general conflict was markedly higher among people who weren't affiliated with a church.

One reason why some churches are eager to embrace science is because they're losing adherents, particularly young ones who feel that Christianity isn't friendly to science. A study by the Barna Group, a market research firm that studies religious issues, found that this is one of six reasons why young folk are abandoning Christianity:

Reason #3—Churches come across as antagonistic to science. One of the reasons young adults feel disconnected from church or from faith is the tension they feel between Christianity and science. The most common of the perceptions in this arena is "Christians are too confident they know all the answers" (35%). Three out of ten young adults with a Christian background feel that "churches are out of step with the scientific world we live in" (29%). Another one-quarter embrace the perception that "Christianity is anti-science" (25%). And nearly the same proportion (23%) said they have "been turned off by the creation-versus-evolution debate." Furthermore, the research shows that many science-minded young Christians are struggling to find ways of staying faithful to their beliefs and to their professional calling in science-related industries.

If the incompatibility of science and religion is an illusion, it's one that's powerful enough to make these young Christians vote with their feet. They may not abandon religion, but they certainly break ties with their church.

While some liberal churches deal with the conflict by simply accepting the science and modifying their theology where required, more conservative ones put up a fight. One of the more remarkable demonstrations of this resistance occurred in September 2013, when a group of parents, with the help of a conservative legal institute, filed suit against the Kansas State

Board of Education. Their goal was to overturn *the entire set of state science standards from kindergarten through twelfth grade*, arguing that those standards gave students a "materialistic atheistic" worldview that was inimical to their religion. Just as this book went to press, the lawsuit was dismissed.

Finally, if religion and science get along so well, why are so many scientists nonbelievers? The difference in religiosity between the American public and American scientists is profound, persistent, and well documented. Further, the more accomplished the scientist, the greater the likelihood that he or she is a nonbeliever. Surveying American scientists as a whole, Pew Research showed that 33 percent admitted belief in God, while 41 percent were atheists (the rest either didn't answer, didn't know, or believed in a "universal spirit or higher power"). In contrast, belief in God among the general public ran at 83 percent and atheism at only 4 percent. In other words, scientists are ten times more likely to be atheists than are other Americans. This disparity has persisted for over eighty years of polling.

When one moves to scientists working at a group of "elite" research universities, the difference is even more dramatic, with just over 62 percent being either atheist or agnostic, and only 23 percent believing in God—a degree of nonbelief more than fifteenfold higher than among the general public.

Sitting at the top tier of American science are the members of the National Academy of Sciences, an honorary organization that elects only the most accomplished researchers in the United States. And here nonbelief is the rule: 93 percent of the members are atheists or agnostics, with only 7 percent believing in a personal god. This is almost the exact opposite of the data for "average" Americans.

Why do so many scientists reject religion compared with the general public? Any answer must also explain the observation that the better the scientist, the greater the likelihood of atheism. Three explanations come to mind. One has nothing to do with science per se: scientists are simply more educated than the average American, and religiosity simply declines with education.

While that is indeed the case, we can rule it out as the only explanation from a 2006 survey of religious belief of university professors in different fields. As with scientists, American university professors were more atheistic or agnostic than the general populace (23 percent versus 7 percent non-believers, respectively). But when professors from different areas were polled, it became clear that scientists were the *least* religious. While only 6 percent of "health" professors were atheists or agnostics, this figure was 29 percent for humanities, 33 percent for computer science and engineering, 39 percent for social sciences, and a whopping 52 percent for physical and biological scientists together. When disciplines were divided more finely, biologists and psychologists tied as the least religious: 61 percent of each group were agnostics or atheists. So, among academics with roughly equal amounts of higher education, scientists still reject God more often. The tentative conclusion is that the atheism of scientists doesn't simply reflect their higher education, but is somehow inherent in their discipline.

That leaves two explanations for the atheism of scientists, both connected with science itself. Either nonbelievers are drawn to become scientists, or doing science promotes the rejection of religion. (Both, of course, can be true.) Accommodationists prefer the first explanation because the latter implies that science itself produces atheism—a view that liberal believers abhor. Yet there are two lines of evidence that practicing science does erode belief. The first is that elite scientists were raised in religious homes nearly as often as nonscientists, yet the former still wind up being far less religious. But this may mean only that religious homes can produce nonbelievers, who then are preferentially drawn to science.

But there's further evidence. If you survey American scientists of different ages, you find that the older ones are significantly less religious than the younger. While this suggests that the erosion of faith is proportional to one's tenure as a scientist, there's an alternative explanation: a "cohort effect." Perhaps older scientists were simply born in an era when religious belief was less pervasive, and have retained their youthful unbelief. But that seems unlikely, for the trend is actually in the opposite direction: the religiosity of Americans has declined over the last sixty years. The "cohort hypothesis" predicts that older scientists would be *more* religious, and they're not.

All of this suggests that lack of religious belief is a side effect of doing science. And as repugnant as that is to many, it's really no surprise. For some people, at least, science's habit of requiring evidence for belief, combined with its culture of pervasive doubt and questioning, must often carry over to other aspects of one's life—including the possibility of religious faith.

In chapter 3 I'll argue that the existence of religious scientists does not constitute strong evidence for the compatibility of science and faith. Isn't it then hypocritical to argue that the existence of atheistic scientists is evidence for an *incompatibility* between science and faith? My response is that religious scientists are in some ways like the many smokers who don't get lung cancer. Just as those cancer-free individuals don't invalidate the statistical relationship between smoking and the disease, so the existence of religious scientists doesn't refute an antagonistic relationship between science and faith. Scientists of faith happen to be the ones who can compartmentalize two incompatible worldviews in their heads.

On the whole, it's difficult to escape the conclusion—based on the paucity of religious scientists, the incessant stream of books using contradictory arguments to promote accommodationism, the constant reassurance by scientific organizations that believers can accept science without violating their faith, and the pervasiveness of creationism in many countries—that there is a problem in harmonizing science and religion, one that worries both sides (but mostly the religious).

After a period of relative quiescence since the books of Draper and White, why has the issue of science versus religion been revived? I see three reasons: recent advances in science that have pushed back the claims of religion, the rise of the Templeton Foundation as a major funder of accommodationist ventures, and, finally, the appearance of New Atheism and its explicit connection with science, especially evolution.

The deadliest blow ever struck by science against faith was Darwin's publication of *On the Origin of Species*. But that was in 1859. The conflict between religion and evolution didn't really get going until religious fundamentalism arose in early-twentieth-century America. An organized push for creationism began around 1960, and then, after a series of court cases prohibiting its teaching in public schools, creationism assumed the guise of science itself—first as the oxymoronic "scientific creationism," pretending that the Bible supported the very facts of science. When that failed, creationism turned into "intelligent design" (ID), whose teaching was also struck down by the courts in 2005. With the failure of ID, which is about as watered down as creation-ism can get, those who reject evolution have become more defensive and vo-ciferous, eager to find other ways to go after science. Ironically, as the credibility of creationists grows smaller, their voices get louder.

In contrast, evolution goes from strength to strength, as new data from the fossil record, molecular biology, and biogeography continue to affirm its hegemony as the central organizing principle of biology. Creationists wait-ing for the decisive evidence against evolution, evidence that ID promised to deliver, have been disappointed. As I said in my previous book, "Despite a million chances to be wrong, evolution always comes up right. That is as close as we can get to a scientific truth." And now the new field of evolution-ary psychology, by studying the evolutionary roots of human behavior, gradually erodes the uniqueness of many human traits, like morality, once imputed to God. As I'll discuss in chapter 4, we see in our evolutionary rel-atives behaviors that look very much like rudimentary morality. This sug-gests that many of our "moral" feelings could be the result of evolution, while the rest could result from purely secular considerations.

Recent advances in neuroscience, physics, cosmology, and psychology have also replaced supernatural explanations with naturalistic ones. Al-though our knowledge of the brain is still scanty, we're beginning to learn that "consciousness," once attributed to God, is a product of diffuse brain activity and not some metaphysical "I" sitting inside our skulls. It can be manipulated and altered with surgery and chemicals, making it a phenom-enon that is surely a product of brain activity. The notion of "free will"—a linchpin of many faiths—now looks increasingly dubious as scientists not only untangle the influence of our genes and environments on our behavior, but also show that some "decisions" can be predicted from brain scans sev-eral seconds before people are conscious of having made them. In other words, the notion of pure "free will," the idea that in any situation we can choose to behave in different ways, is vanishing. Most scientists and philos-ophers are now physical "determinists" who see our genetic makeup and environmental history as the only factors that, acting through the laws of physics, determine which decisions we make. That, of course, kicks the props out from under much theology, including the doctrine of salvation through freely choosing a savior, and the argument that human-caused evil is the undesirable but inevitable by-product of the free will vouchsafed us by God.

In physics, we are starting to see how the universe could arise from "nothing," and that our own universe might be only one of many universes that differ in their physical laws. Far from making us the special objects of God's attention, such a cosmology sees us simply as holders of a winning lottery ticket—the inhabitants of a universe that had the right physical laws to allow evolution.

Bit by bit, the list of phenomena that once demanded an explanatory God is being whittled down to nothing. Religion's response has been to either reject the science (the tactic of fundamentalists) or bend their theology to accommodate it. But theology can be bent only so far before, by rejecting theological nonnegotiables like the divinity of Jesus, it snaps, turning into nonreligious secular humanism.

That gives another clue to the rise in accommodationism, at least in America: the recent decline in formal religious affiliation. The percentage of Americans who either are nonbelievers or claim no religious affiliation— the so-called nones—is rising rapidly. The proportion of atheists, agnostics, and those who are spiritual but not religious stood at 20 percent in 2012, up 5 percent from 2005. This makes "nones" the fastest-growing category of "believers" in America. This trend is well known and recognized by the churches, and, as we've seen, partly reflects how young people are turned off by religion's perceived antagonism to science.

How can religion stem this attrition? For those who want to keep the comforts of their faith but not appear backward or uneducated, there is no choice but to find some rapport between religion and science. Besides trying to retain adherents, churches have a further reason to embrace science: liberal theology prides itself on modernism, and there is no better way to profess modernity than to embellish your theology with science. Finally, everyone, including believers, recognizes the remarkable improvements in our quality of life over the past few centuries, not to mention remarkable technical achievements like sending space probes to distant planets. And everyone knows that those achievements come from science, its ability to find the truth and then to use those findings to promote not only further understanding but improvements in technology and human well-being. If you see your religion as also making salubrious claims about the truth, then you must recognize that it is in some ways competing with science—and not too successfully. After all, what new insights has religion produced in the last century? This disparity in outcome might well cause some cognitive dissonance, a mental discomfort that can be resolved—though not very well—by arguing that there's no conflict between science and religion.

Much of the recent spurt of accommodationism has been fueled by the funds of a single organization—the John Templeton Foundation. Templeton (1912–2008) was a billionaire mutual-fund magnate knighted by Queen Elizabeth after he moved to the Bahamas as a tax exile. Although a Presbyterian, he was convinced that other religions also held clues to "spiritual" realities and that, indeed, science and religion could be partners in solving the "big questions" of purpose, meaning, and values. To that end he bequeathed his fortune—the endowment is now $1.5 billion—to his eponymous foundation, set up in 1987. Its philanthropic mission reflects Templeton's push for accommodationism:

> Sir John believed that continued scientific progress was essential, not only to provide material benefits to humanity but also to reveal and illuminate God's divine plan for the universe, of which we are a part.

The foundation's main philanthropic goal is funding work on what it calls "the Big Questions": areas that clearly mix science with religion. As the foundation states:

> Sir John's own eclectic list featured a range of fundamental scientific notions, including complexity, emergence, evolution, infinity, and time. In the moral and spiritual sphere, his interests extended to such basic phenomena as altruism, creativity, free will, generosity, gratitude, intellect, love, prayer, and purpose. These diverse, far-reaching topics define the boundaries of the ambitious agenda that we call the Big Questions. Sir John was confident that, over time, the serious investigation of these subjects would lead humankind ever closer to truths that transcend the particulars of nation, ethnicity, creed, and circumstance.

. . . For Sir John, the overarching goal of asking the Big Questions was to discover what he called "new spiritual information." This term, to his mind, encompassed progress not only in our conception of religious truths but also in our understanding of the deepest realities of human nature and the physical world. As he wrote in the Foundation's charter, he wanted to encourage every sort of opinion leader—from scientists and journalists to clergy and theologians—to become more open-minded about the possible character of ultimate reality and the divine.

The Templeton Foundation distributes $70 million yearly in grants and fellowships. To put that in perspective, that's *five times* the amount dispensed annually by the U.S. National Science Foundation for research in evolutionary biology, one of Templeton's areas of focus. Given Templeton's deep pockets and not overly stringent criteria for dispensing money, it's no wonder that, in a time of reduced financial support, scientists line up for Templeton grants.

And from that support flows a constant stream of conferences, books, papers, and magazine articles, many arguing for harmony between faith and science. You may have encountered the foundation through its full-page ads in the *New York Times*, with noted scholars (many already supported by Templeton) discussing questions like "Does science make belief in God obsolete?" and "Does the universe have a purpose?"

The foundation's most famous award, originally named the Templeton Prize for Progress in Religion, is now called simply the Templeton Prize, given to "[honor] a living person who has made an exceptional contribution to affirming life's spiritual dimension, whether through insight, discovery, or practical works." It goes to a single individual, who gets £1.1 million (roughly $1.8 million), an amount deliberately set to exceed that of the Nobel Prize (about $1.2 million, shareable by up to three recipients). Originally awarded to religious figures, theologians, and philosophers, including Billy Graham, Mother Teresa, and Watergate conspirator Charles Colson, the prize is now also given to religion-friendly scientists like the evolutionary biologist Francisco Ayala and the cosmologist Martin Rees.

The change in both the prize's name and the nature of its recipients suggests an increasing desire of the foundation to downplay its religious side and acquire more of the cachet of science. After all, many scientists are reluctant to engage with explicitly religious organizations. Indeed, Templeton does fund some pure science. But its mission, to promote accommodationism, has remained the same, and the funded projects nearly always have a theological side. Templeton, for instance, funded the Faraday Institute for Science and Religion at Cambridge, which also runs a program for children, "Test of Faith," showing how Christianity comports with science. Templeton funds the BioLogos Foundation, which is designed to show evangelical Christians that they can accept both Jesus and Darwin. Templeton funds the Dialogue on Science, Ethics, and Religion program of the American Association for the Advancement of Science, a curiously theological arm of a large scientific organization, one designed to promote the idea that the relationship between science and religion is purely positive.

Templeton's time-limited "research grants" are often hybrids between science and religion. There is, for instance, a three-year, $5.1 million "Immortality Project," devoted to studying the afterlife, its possible manifestation through near-death experiences, its influence on people's behavior, and its characteristics. Such a mixture requires not only the labor of sociologists, but also the lucubrations of theologians. Templeton awarded $5.3 million for a project called "The Science of Intellectual Humility," which is heavy on theology and light on science. It gave $1.7 million for the project "Randomness and Divine Providence," with a combination of physicists, mathematicians, and theologians studying how randomness in nature might be consistent with the existence of a loving god. And $4.4 million went for the study "Big Questions in Free Will," involving philosophers, neuroscientists, and, of course, theologians.

One of Templeton's biggest grants—$10.5 million over five years—was awarded to a group of scientists in my own field: a study titled "Foundational Questions in Evolutionary Biology" led by Martin Nowak at Harvard University. And while some of that money was directed toward valid scientific questions, like studying the conditions that promote the evolution of cooperation, the project included some distinctly nonscientific components:

> The Foundational Questions in Evolutionary Biology initiative at Harvard
> University seeks to generate new kinds of knowledge and understanding
> in core areas of biology. FQEB encourages researchers to explore such

topics as the origins of biological creativity, the deep logics of biological dynamics and biological ontology, and concepts of teleology and ultimate purpose in the context of evolution. Such knowledge is directly relevant to a wide range of philosophical and theological discussions and debates.

The notions of ultimate purpose and "teleology" (an external force directing evolution) are simply not part of science: this mixing of the scientific with the metaphysical is characteristic of Templeton's approach. One would think that scientists would be wary of participating in programs that dilute and even distort science in this way, but one would underestimate scientists' need for research money. And Templeton benefits as well, for the funded scientists are paraded on its Web site like prize horses, evidence of serious purpose and of a fruitful dialogue between science and faith.

Besides funding science and accommodationism, Templeton also gives money to purely religious projects, such as the television show *The American Bible Challenge* and the $100,000 Epiphany Prize awarded for "the best wholesome, uplifting and inspiring movies and television programs." (That prize was once awarded to the gruesome and anti-Semitic movie *The Passion of the Christ*.) Templeton also gave a $3 million grant to Biola University (formerly the Bible Institute of Los Angeles), an evangelical Christian school in California, to found a Center for Christian Thought. It was the largest foundation grant in the school's history.

Given the eagerness of many cash-strapped scientists to join the Templeton stable, the influence of its money on the syncretic program of science and faith should not be underestimated. Whenever you see a public discussion of science and faith, at least in the United States, chances are there's Templeton money behind it. Even the World Science Festival, run by the physicist Brian Greene and his partner Tracy Day, was partly founded with Templeton money. And so, along with the many lectures and demonstrations of science, there is always a session on "Big Ideas," often featuring Templeton Prize winners and discussions of science and metaphysics.

Finally, it's clear that accommodationism in the last decade has been partly a response to the popularity of New Atheism and its dissemination via the Internet and several best-selling books. Important New Atheist works include *The End of Faith* and *Letter to a Christian Nation*, by Sam Harris; *The God Delusion*, by Richard Dawkins; *Breaking the Spell*, by Daniel Dennett; *God Is Not Great: How Religion Poisons Everything*, by Christopher Hitchens; and *God: The Failed Hypothesis; How Science Shows That God Does Not Exist*, by Victor Stenger. All these books appeared within a span of three years, and all were *New York Times* bestsellers. Although they reprised much of the "old atheism" of people like Robert Ingersoll, Bertrand Russell, and H. L. Mencken, bringing forgotten arguments to a new generation, there was also a new element: an explicit connection with science. More than ever before, atheists with a public profile come from the ranks of scientists and science aficionados. Dawkins is an evolutionary biologist, Stenger (recently deceased) was a physicist, Harris is a trained neuroscientist, and Dennett is a philosopher of science and the mind. Hitchens, a journalist, is the sole exception, but even he was widely read in science and frequently dealt with evolution and cosmology.

Nevertheless, these books were more concerned with showing the inimical effects of faith and refuting its claims than with exploring the complex relationship between science and religion. Now that the dust is settled, it's time to examine in detail why science and religion are incompatible, and to scrutinize the common arguments for their compatibility. And it's time to address the new "natural theological" arguments for God: the supposed "fine-tuning" of physical laws, the existence of supposedly innate moral sentiments, and other areas where religion equates scientific ignorance with evidence for God. Finally, we need to determine whether there's any benefit to a dialogue between scientists and believers—the kind of dialogue that the Templeton Foundation promotes with its constant infusions of cash.

The scientific bent of New Atheism, and the issue on which I center this book, is reflected in its view that *religious claims are empirical hypotheses.* This is not a profound realization. After all, it's palpably clear that most religions make claims about what is true in our universe—that is, empirical claims. Here are some examples from just one faith, Trinitarian Christianity: There is a God who intercedes in the affairs of humans. God created humans in his image but then two of them sinned, infecting all of their

descendants—the entire species of *Homo sapiens*—with a taint that did not
exist before. The deity also fathered a son by a virgin female, a son whose
execution and Resurrection gave us the opportunity to expiate our inherited sins. Further, there is an afterlife in which those who were virtuous in
their earthly lives will dwell in paradise, while miscreants suffer eternity in
hell. Only those who accept Jesus as savior will enjoy the delights of heaven
("I am the way, the truth, and the life: no man cometh unto the Father, but
by me"). Finally, Jesus will return someday, ushering in the final reckoning
of the End Times. And prayer can work: God listens to our supplications
and sometimes grants them.

These aren't just the claims of fundamentalists. As we'll see in the next
chapter, far more than half of *all* Americans take them literally. Catholicism, with its strict moral code, goes further. Acts like unconfessed mastur-
bation, homosexual sex, adultery, and so on are explicitly classified as grave
sins, punishable by an eternity in hell. Granted, not every Catholic agrees
with this, but every theist—and most believers are theists—believes that
God interacts with the world in *some* way.

And, of course, such claims aren't unique to Christianity. Islam has its
own God, heaven, and hell, Judaism its Messiah, whose return, hastened by
acts of goodness and piety, is anticipated within the next few centuries. Christian Science sees diseases not as organic ailments, but as a result of improper
thinking—a belief shared by many Pentecostal Christian sects. Some Buddhist sects, and many Hindus, believe in reincarnation, and Buddhism adds
karma, a doctrine of cosmic retribution for good and bad actions. Scientology
(whose "theology" seems bizarre simply because we were around when it was
invented) uses an e-meter, a device that measures electrical current across the
skin, in "auditing" sessions designed to clear the body of malevolent spirits
("thetans") that supposedly afflict many of us. An important claim of Scientology is that psychiatry and its medications are scams that have no beneficial
effect. The cargo cults of Melanesia, like the famous cult of "John Frum," assume that by building replicas of airplanes and airports and worshipping
Frum (whose origin is unknown, but apparently stands for American GIs who
brought goods to the islands in World War II), they can acquire more of the
goods their relatives got seventy years ago. Needless to say, the swag never
arrives.

I could go on, but the point is clear: religions make explicit claims about reality—about what exists and happens in the universe. These claims involve the existence of gods, the number of such gods (polytheism or monotheism), their character and behavior (usually loving and beneficent, but, in the case of Hindu and ancient Greek gods, sometimes mischievous or malevolent), how they interact with the world, whether or not there are souls or life after death, and, above all, how the deities wish us to behave—their moral code.

These are empirical claims, and although some may be hard to test, they must, like all claims about reality, be defended with a combination of evidence and reason. If we find no credible evidence, no good reasons to believe, then those claims should be disregarded, just as most of us ignore claims about ESP, astrology, and alien abduction. After all, beliefs important enough to affect you for eternity surely deserve the closest scrutiny. Christopher Hitchens was fond of saying that "extraordinary claims require extraordinary evidence." His inevitable corollary was that "what can be asserted without evidence can also be dismissed without evidence." The philosopher L. R. Hamelin describes what happens when we apply science to the existence of God, stipulating five criteria for the "God theory":

First, we hypothesize that God is *real*, with real properties. Second, we create a theory about what a real God and His properties means. A God doesn't just sit there; what does He *do*? Third, we make this theory *testable:* we must be able to determine whether it is true or false. Fourth, we must test the theory by observation or experiment. Finally, we ensure the theory is parsimonious: that is, if we took out God, the theory wouldn't explain as much. Once we have followed all these steps, we have a scientific theory that includes God, which we can test against what we actually observe.

But constructing this kind of theory of God puts believers on the horns of a dilemma. Centuries of scientific investigation show that the best scientific theories, testable by observation, include nothing like a personal God. We find only a universe of blind, mechanical laws, including natural selection, with no foresight or ultimate purpose.

Alternatively, a believer could reject one or more of the criteria for a

God theory, but doing that has profound implications. If she admits that God is not real, she's already an atheist. If she says God doesn't do anything, who cares? If her theory cannot be tested at all, then there's no way of telling if it's true or false. If her theory can be tested only by private revelation, not by observations available to everyone, she unjustifiably claims private knowledge. And if her theory is observationally identical to a theory that does not include God, then she's again an atheist, for a God who makes no difference is no God at all.

The only remaining question is whether some people would find this analysis useful, and I know many people who, applying this analysis, have abandoned their religion.

Some may find this excessively philosophical, saying that their belief in God rests not on logic but on emotions. But all it does is formalize the criteria we use for accepting anything as "real."

Throughout this book we'll encounter not only the kinds of empirical claims made by the faithful, but also the ways that different religions make different and contradictory claims—claims that have often led to religious schisms. Like the branching tree of biological life itself, religions have splintered and proliferated in ways that have created new sects that can be placed in a genealogy of faith. Just as there are millions of species, so there are thousands of faiths. The diversity of their conflicting claims suggests that we should be skeptical about the tenets of every faith. But we'll also examine the ways that believers, knowing the fragility of such claims, try to insulate themselves from having their empirical claims tested—or even discussed. The ways of rejecting the call for evidence include denying that faith *needs* evidence, arguing that religion is really an exercise in metaphor, a series of parables that aren't meant to be historically accurate. Alternatively, some argue that religion makes no existence claims at all. That last category includes adherents to apophatic theology, which says that one can say nothing about the nature of God (although his existence never seems to be in question, and the books that say nothing about him are many), as well as those who assert that God isn't a humanlike spirit, but a nebulous "ground of being" that defies concrete description. Finally, while many believers admit that religion does make existence claims, they argue that those claims involve "ways of knowing" that bypass reason and evidence. Those ways involve private revelation, church authority, and, especially, faith—the acceptance of things for which there is no strong evidence.

It does not seem misguided to me, or an insult to believers, to regard God and much religious dogma as hypotheses. As we'll see in the next chapter, religions regularly make statements about what exists in the cosmos, and believers' acceptance of those statements—as well as their rejection of empirical claims made by other religions—ultimately depends on what they consider to be *evidence*. Is it disrespectful to take the claims of believers seriously and examine them rationally and scientifically, especially when so much of modern society, including law, politics, and morality, rests on those claims?

If there is a common theme in the writings of those who see religion and science as incompatible, it is the idea that *in science faith is a vice, while in religion it's a virtue*. It is this incompatibility that upsets believers when skeptics use reason to scrutinize the tenets of their faith—whether those tenets involve the Resurrection, the authenticity of the Book of Mormon, or the reward of a coterie of virgins in paradise for religious martyrdom. The rational scrutiny of religious faith involves asking believers only two questions:

How do you know that?

What makes you so sure that the claims of your faith are right and the claims of other faiths are wrong?

I've argued that there is plenty of evidence, some involving people's perception, that the purported harmony between science and religion is not what it's cracked up to be. In the next chapter, I'll go beyond perception and give my own explanation for why these areas are irredeemably incompatible.

CHAPTER 2

What's Incompatible?

I admit I'm surprised whenever I encounter a religious scientist. How can a bench-hazed Ph.D., who might in an afternoon deftly purée a colleague's PowerPoint presentation on the nematode genome into so much fish chow, then go home, read in a two-thousand-year-old chronicle, riddled with internal contradictions, of a meta-Nobel discovery like "Resurrection from the Dead," and say, gee, that sounds convincing? Doesn't the good doctor wonder what the control group looked like?

—Natalie Angier

I learned about the nature of science the hard way. After an undergraduate education in biology at a small southern college, I was determined to get a Ph.D. in evolutionary genetics at the best laboratory in that field. At the time, that was the laboratory of Richard Lewontin at Harvard's Museum of Comparative Zoology, for Lewontin was widely seen as the world's best evolutionary geneticist. But soon after I arrived and began working on evolution in fruit flies, I thought I'd made a terrible mistake.

Shy and reserved, I felt as if I'd been hurled into a pit of unrelenting negativity. In research seminars, the audience seemed determined to dismantle the credibility of the speaker. Sometimes they wouldn't even wait until the question period after the talk, but would rudely shout out critical questions and comments during the talk itself. When I thought I had a good idea and tentatively described it to my fellow graduate students, it was picked apart like a flounder on a plate. And when we all discussed science around the big rectangular table in our commons room, the atmosphere was heated and contentious. Every piece of work, published or otherwise, was scrutinized for problems—problems that were almost always found. This made me worry that whatever science I managed to produce could never make the grade. I even thought about leaving graduate school. Eventually, fearful of being criticized, I simply kept my mouth shut and listened. That went on for two years.

But in the end, that listening was my education in science, for I learned that the pervasive doubt and criticality weren't intended as personal attacks, but were actually the essential ingredients in science, used as a form of quality control to uncover the researcher's misconceptions and mistakes. Like Michelangelo's sculpturing, which he saw as eliminating marble to reveal the statue within, the critical scrutiny of scientific ideas and experiments is designed, by eliminating error, to find the core of truth in an idea. Once I'd learned this, and developed a skin thick enough to engage in the inevitable to-and-fro, I began to enjoy science. For if you can tolerate the criticality and doubt—and they're not for everyone—the process of science yields a joy that no other job confers: the chance to be the first person to find out something new about the universe.

Until I started pondering the relationship of science and religion for this book, I never really thought about what "science" was, although I'd been doing it for over three decades. Most scientists never get formal instruction in "the scientific method," except perhaps for the rote (and incorrect) recitation of "make hypothesis/test it/accept it" sequence you see in textbooks. Literally and figuratively, I learned science on the fly, simply by watching how my peers did it. But learning it and defining it are different matters. In fact, it was not until I wrote this book that I realized that my own notion of science is simply that of a method: a process (to my mind, the *only* process) that has proved useful in helping us understand what is real in the universe. While I had never pondered this issue, my training as a scientist had led me to unconsciously internalize its methods.

What Is Science?

So what is "science"? Before I lay out my own definition, let's see how the word is construed by other people. To many it represents simply the *activities* of professional scientists: the person on television in the lab coat who, peddling the latest antiwrinkle cream, touts it with the words "Science says ..." To others, it's the *knowledge* produced by scientists: the facts taught in classes on chemistry, biology, physics, geology, and so on. Those facts segue into *technology,* or the practical applications of scientific knowledge—the development of antibiotics, computers, lasers, and so on.

But scientific knowledge is often transitory: some (but not all) of what we find is eventually made obsolete, or even falsified, by new findings. That is not a weakness but a strength, for our best understanding of phenomena will alter with changes in our way of thinking, our tools for looking at nature, and what we find in nature itself. Any "knowledge" incapable of being revised with advances in data and human thinking does not deserve the name of knowledge. In my lifetime, the continents were thought to be static, but now we know they move—at the same rate our fingernails grow. The universe was also thought by many to be static, having eternally been in its present form, until 1929, when Edwin Hubble showed that it was expanding, and later, in 1964, when scientists discovered the background radiation that was the sign of a Big Bang. And even in 1949, the year I was born, and less than three years before Watson and Crick discovered the structure of DNA, many people still thought that the genetic material was a protein.

What is "known" may sometimes change, so science isn't really a fixed body of knowledge. What remains is what I really see as "science," which is simply *a method for understanding how the universe (matter, our bodies and behavior, the cosmos, and so on) actually works.* Science is a set of tools, refined over hundreds of years, for getting answers about nature. It is the set of methods we cite when we're asked "How do you *know* that?" after making claims such as "Birds evolved from dinosaurs" or "The genetic material is not a protein, but a nucleic acid."

My view of science as a toolkit is what Michael Shermer meant when he defined science as a collection of methods that produce "a testable body of knowledge open to rejection or confirmation." That's as good a definition of science as any, but the best *rationale* for using those methods came from the renowned and colorful physicist Richard Feynman:

The first principle is that you must not fool yourself—and you are the easiest person to fool. So you have to be very careful about that.

Like everyone, scientists can suffer from *confirmation bias,* our tendency to pay attention to data that confirm our a priori beliefs and wishes, and to ignore data we don't like. But, like all rational people, we must admit the truth of what Voltaire noted in 1763: "The interest I have in believing in something is not a proof that the something exists." The doubt and criticality of science are there precisely for the reason Feynman emphasized: to prevent us from believing what we'd like to be true. The part of Feynman's quote about fooling yourself is important, because his view of science is precisely the opposite of how religion finds truth. (Feynman was an atheist, and I can't help but suspect that he was thinking of religion when he wrote that.) As we'll see, religion is heavily laden with the kind of confirmation bias that makes people see their own faiths as true and all others as false. In other words, religion is replete with features to *help* people fool themselves.

But I'm getting ahead of myself. When I characterize science as a way to find truth, what I mean is "truth about the universe"—the kind of truth that is defined by the *Oxford English Dictionary* as "Conformity with fact; agreement with reality; accuracy, correctness, verity (of statement or thought)." And if you look up "fact," you'll find that it's defined as "something that has really occurred or is actually the case; something certainly known to be of this character; hence, a particular truth known by actual observation or authentic testimony, as opposed to what is merely inferred, or to a conjecture or fiction; a datum of experience, as distinguished from the conclusions that may be based upon it."

In other words, truth is simply what *is:* what exists in reality and can be verified by rational and independent observers. It is true that DNA is a double helix, that the continents move, and that the Earth revolves around the Sun. It is not true, at least in the dictionary sense, that somebody had a revelation from God. The scientific claims can be corroborated by anyone with the right tools, while a revelation, though perhaps reflecting someone's real *perception,* says nothing about reality, for unless that revelation has empirical content, it cannot be corroborated. In this book I will avoid the murky waters of epistemology by simply using the words "truth" and "fact" interchangeably. These notions blend into the concept of "knowledge," defined as "the apprehension of fact or truth with the mind; clear and certain perception of fact or truth; the state or condition of knowing fact or truth."

As I noted above, widespread agreement by scientists about what is true does not guarantee that that truth will never change. Scientific truth is never absolute, but provisional: there is no bell that rings when you're doing science to let you know that you've finally reached the absolute and unchangeable truth and need go no further. Absolute and unalterable truth is for mathematics and logic, not empirically based science. As the philosopher Walter Kaufmann explained, "What distinguishes knowledge is not certainty but evidence."

And that evidence can change. It's easy to find cases of accepted scientific "truths" that were later shown to be false. I've mentioned a few above, and there are many more. Early cases in the history of science are geocentrism (the Earth as the center of the cosmos) and the Greek concept of the "four humors": that both personality and disease resulted from the balance of four bodily fluids (black bile, yellow bile, phlegm, and blood). A famous modern case is the demonstration of "N rays," a form of radiation described in 1903, observed by many people, and then found to be bogus, a result of confirmation bias. Atoms were once considered indivisible particles of matter. There's even one case of a Nobel Prize awarded for a bogus discovery, that of the *Spiroptera carcinoma*, a parasitic nematode worm that supposedly caused cancer. Its discovery earned Johannes Fibiger the Nobel Prize in Physiology or Medicine in 1926. Soon thereafter, researchers showed that this result was wrong: the worm was simply an irritant that, like many other factors, induced tumors in already damaged cells. But Fibiger's prize stands, for his discovery seemed true at the time.

The overturning of some scientific truths has often served as ammunition for religious critics who indict the field for its inconstancy. Science can be *wrong*! But that mischaracterizes *any* attempt to understand truth, both religious and scientific. Scientific tools and ways of thinking change: how can our understanding of nature not change as well? And, of course, the criticism of inconstancy can be turned right back on religion. There is simply no way that *any* faith can prove beyond question that its claims are true while those of other faiths are false.

It is a common saying among scientists that we can prove theories wrong (it would be relatively easy to show, for instance, that the formula for water isn't H_2O), but that we can never prove them *right*, for new observations could always come along that would overturn received knowledge. The theory of evolution, for instance, is regarded by all rational scientists as true, as it's supported by mountains of evidence from many different fields. Yet there are observations that could, if they surfaced, conceivably disprove that theory. These include, for instance, finding fossils embedded in strata from the "wrong" time, like discovering mammalian fossils in four-hundred-million-year-old sediments, or observing adaptations in one species that are useful only for another species, such as a pouch on a wallaby that can hold only baby koalas. Needless to say, such evidence hasn't appeared. Evolution, then, is a fact in the scientific sense, something Steve Gould defined as an observation "confirmed to such a degree that it would be perverse to withhold provisional assent." Indeed, the only real "proofs" beyond revision are those found in mathematics and logic.

But some people take this too far, claiming that scientific truths not only are provisional, but change *most of the time*. Science, the argument goes, isn't really *that* good at apprehending truth, and we should be wary of it. Such claims of inconstancy usually involve medical studies—like the value of a daily aspirin in preventing heart disease, or the advisability of annual mammograms—whose conclusions go back and forth when different populations are sampled. What's important to remember is that most scientific findings become truths when they're replicated many times, either directly by other dubious scientists or when repeated as a foundation for further work.

In reality, we can consider many scientific truths to be about as absolute as truths can be, ones that are very unlikely to change. I would bet my life savings that the DNA in my cells forms a double helix, that a normal water molecule has two hydrogen atoms and one oxygen atom, that the speed of light in a vacuum is unchanging (and close to 186,000 miles per second), and that the closest living relatives of humans are the two species of chimpanzees. After all, you bet your life on science every time you take medicines like antibiotics, insulin, and anticholesterol drugs. If we consider "proof" in the vernacular to mean "evidence so strong that you'd bet your house on it," then, yes, science is sometimes in the business of proof.

So what are the components of the toolkit of science? Like many of us, I was taught in high school that there is indeed a "scientific method," one consisting of "hypothesis, test, and confirmation." You made a hypothesis (for instance, that DNA is the genetic material) and then tested it with laboratory experiments (the classic one, done in 1944, involved inserting the DNA of a disease-causing bacteria into a benign one and seeing if the transformed bacteria could both cause disease and pass this pathogenicity on to its descendants). If your predictions worked, you had supported your hypothesis. With strong and repeated support, the hypothesis was finally considered "true."

But scientists and philosophers now agree that there is no single scientific method. Often you must gather facts before you can even *form* a hypothesis. One example is Darwin's observation, made on his *Beagle* voyage, that oceanic islands—usually volcanic islands that rose above the sea bereft of life—have lots of birds, insects, and plants that are *endemic*, native only to those islands. The diverse species of finches of the Galápagos and the fruit flies of Hawaii are examples. Further, oceanic islands like Hawaii and the Galápagos either have very few species of native reptiles, amphibians, and mammals or lack them completely, yet such creatures are widely distributed on continents and "continental islands" like Great Britain that were once connected to major landmasses. It is these facts that helped Darwin concoct the theory of evolution, for those observations can't be explained by creationism (a creator could have put animals wherever he wanted). Rather, they lead us to conclude that endemic birds, insects, and plants on oceanic islands descended, via evolution, from ancestors that had the ability to migrate to those places. Insects, plant seeds, and birds can colonize distant islands by flying, floating, or being borne by the wind, while this is not possible for mammals, reptiles, and amphibians. Collecting that data and then recognizing a pattern in it was what helped produce the theory of evolution.

And sometimes the "tests" of hypotheses don't involve experiments, but rather observations—often of things that occurred long ago. It's hard to do experiments about cosmology, but we're completely confident in the existence of the Big Bang because we observe things predicted by it, like the expanding universe and the background radiation that is the echo of that event. Historical reconstruction is a perfectly valid way of doing science, so long as we can use observations to test our ideas (this, by the way, makes archaeology and history disciplines that are, in principle, scientific). Creationists often criticize evolution because it can't be seen in "real time" (although it has been), apparently ignorant of the massive *historical* evidence, including the fossil record, the useless remnants of ancient DNA in our genome, and the biogeographic pattern I described above. If we accept as true only the things we see happen with our own eyes in our own lifetime, we'd have to regard all of human history as dubious.

While scientific theories can make predictions, they can also be tested by what I call "retrodictions": facts that were previously known but unexplained, and that suddenly make sense when a new theory appears. Einstein's general theory of relativity was able to explain anomalies in the orbit of Mercury that could not be explained by classical Newtonian mechanics. A thick coat of hair, the lanugo, develops in a human fetus at about six months after fertilization but is usually shed before birth. That makes sense only under the theory of evolution: the hair is a vestige of our common ancestry with other primates, who develop the same hair at a similar stage but don't shed it. (A coat of hair is simply not useful for a fetus floating in warm fluid.)

Finally, it's often said that the defining characteristic of science is that it is *quantitative:* it involves numbers, calculations, and measurements. But that too isn't always true. There's not a single equation in Darwin's *On the Origin of Species,* and the whole theory of evolution, though sometimes *tested* quantitatively, can be stated explicitly without any numbers.

As some philosophers have noted, the scientific method boils down to the notion that "anything goes" when you're studying nature—with the proviso that "anything" is limited to combinations of reason, logic, and empirical observation. There are, however, some important features that distinguish science from pseudoscience, from religion, and from what are euphemistically called "other ways of knowing."

Falsifiability via Experiments or Observations

Although philosophers of science argue about its importance, scientists by and large adhere to the criterion of "falsifiability" as an essential way of finding truth. What this means is that for theory or fact to be seen as correct, there must be ways of showing it to be wrong, and those ways must have been tried and have failed. I've mentioned how the theory of evolution is in principle falsifiable: there are dozens of ways to show it wrong, but none have done so. When many attempts to disprove a theory fail, and that theory remains the best explanation for the patterns we see in nature (as is evolution), then we consider it true.

A theory that cannot be shown to be wrong, while it may be *pondered* by scientists, cannot be accepted as scientific truth. When I was a child I made my first theory: that when I left my room, all my plush animals would get up and move around. But to account for the fact that I never actually saw them move or change their positions during my absence, I added a proviso: the animals would instantly assume their former positions when I tried to catch them. At the time, that was an unfalsifiable hypothesis (nanny cams didn't exist). That seems silly, but is not too far removed from theories about paranormal phenomena, whose adherents claim—as they often do for ESP or other psychic "powers"—that the presence of observers actually eliminates the phenomenon. Likewise, claims of supernatural phenomena, such as the efficacy of prayer are rendered unfalsifiable by the assertion that "God will not be tested." (Of course, if the tests had been successful, then testing God would have been fine!) A more scientific example of untestability is that of string theory, a branch of physics claiming that all fundamental particles can be represented as different oscillations on one-dimensional "strings," and that the universe may have twenty-six dimensions instead of four. String theory is enormously promising because if it is right it could constitute the elusive "theory of everything" that unifies all known forces and particles. Alas, nobody has thought of a way of testing it. Absent such tests, it stands as a fruitful theory, but because it's not at present falsifiable, it's one that can't be seen as true. In the end, a theory that can't be shown to be wrong can never be shown to be right.

Doubt and Criticality

Any scientist worth her salt will, when getting an interesting result, ask several questions: Are there alternative explanations for what I found? Is there a flaw in my experimental design? Could anything have gone wrong? The reason we do this is not only to make sure that we have a solid result but also to protect our reputation. There's no better incentive for honesty than the knowledge that you're competing against other scientists in the same area, some often working on the very same problem. If you screw up, you'll be found out very quickly.

That, by the way, gives the lie to the many creationists who claim that we evolutionists conspire to prop up a theory we supposedly know is wrong. They never specify what motivates us to keep promoting something that they consider so obviously false, but creationists often imply that we're committed to using evolution as a way to buttress the atheism of science. (Never mind that many scientists, including evolutionary biologists, are believers, with no vested interest in promoting atheism.) But the main argument against conspiracy theories in science is that anyone who could disprove an important paradigm like the modern theory of evolution would gain immediate renown. Fame accrues to those who, like Einstein and Darwin, overturn the accepted explanations of their day, not to journeymen who simply provide additional evidence for theories that are already widely accepted.

A striking example of the importance of doubt was the finding in 2011 that neutrinos appeared to move faster than the speed of light, discovered by timing their journey over a path from Switzerland to Italy. That observation was remarkable, for it violated everything we know about physics, especially the "law" that nothing can exceed the speed of light. Predictably, the first thing that the physicists (and almost every scientist) thought when hearing this report was simply, "What went wrong?" Although if such an observation were correct it would surely garner a Nobel Prize, one would risk a lifetime of embarrassment to publish it without substantial replication and checking. And, sure enough, immediate checks found that the neutrinos had behaved properly, and their anomalous speed was due simply to a loose cable and a faulty clock.

Replication and Quality Control

Although unique observations (those reported in a single paper) are common in some areas of science, particularly whole-organism biology like evolution and ecology, in most fields, including chemistry, molecular biology, and physics, results are constantly being replicated by other observers. In those areas results become "true" only when they're repeated often enough to gain credibility. The discovery of the Higgs boson in 2012, for which Peter Higgs and François Englert received a Nobel Prize the next year, was deemed prize-worthy because it was confirmed by two completely independent teams of researchers, each using rigorous statistical analysis.

A sufficiently novel or startling result will immediately inspire doubtful scientists to repeat it, often bent on disproving it. Other scientists, assuming your results are correct, might try to build on them to find new things, and part of that involves verifying your original results. The whole edifice of modern molecular genetics depends on the accuracy of the double-helix model of DNA, its process of replicating by unzipping and using each strand as a template to build another, and on the notion that the genetic code involved triplets of bases, each triplet coding for one unit (amino acid) of a protein. If any of this had been wrong, it would have been discovered very quickly as the field advanced. Likewise, each advance tested by proxy all the preceding ones.

Science has additional features that keep us from fooling ourselves by conscious or unconscious finagling with experiments or data. These include statistical analyses that tell us how likely our results might have been due to chance alone rather than to our new theory; blind testing, in which the researcher is prevented from knowing what material she's testing ("double-blind studies," in which neither researcher nor patient knows the identity of the treatment being given, are the gold standard for drug testing); and data sharing, which requires scientists to provide their raw data to anyone who asks, ensuring that those who want to can search for anomalies and run their own statistical tests.

Parsimony

Scientific theories invoke no more factors than necessary to adequately explain any phenomenon. This, like everything in the toolkit of science, is not an a priori requirement of the scientific method, but simply a method developed over centuries of experience. In this case, ignoring things that seem irrelevant keeps us from distracting ourselves with false leads. If we can completely explain the presence of smallpox by infection with a virus, why even *consider* factors like whether the patient ate too much sugar, or, indeed, whether, as was once thought, he was being divinely punished for immorality?

One unparsimonious method is invoking gods. Our experience that su-pernatural hypotheses have never advanced our understanding of the cosmos has, as we'll see later, led to the idea of *philosophical naturalism:* the notion that supernatural entities not only fail to help us understand nature, but don't seem to exist at all.

Living with Uncertainty

One of the most common statements we hear in science is "I don't know." Scientific papers, even those that report fairly solid findings, are hedged with statements like "this suggests that . . . ," or "if this finding is correct . . . ," or "this result should be verified by further experiments." Granted, scientists are people, and we'd like to know all the answers, but in the end it's our ignorance that moves science forward. It's no shame to admit it, for without the unknown, there would be no science, nothing to spark our curiosity. But that attitude assumes that there are some answers we might *never* know.

One of these is how life originated. We know it happened between 4.5 billion years ago, when the Earth was formed, and 3.5 billion years ago, when we already see the first bacterial fossils. And we're virtually certain that all living creatures descended from one original life-form, for virtually all species share the same DNA code, something that would be a remark-able coincidence if the code arose several times independently. But because the first self-replicating organism was small and soft-bodied and thus could not fossilize (it was likely a molecule, perhaps one surrounded by a cell-like membrane), we don't have a way of recovering it.

Now, we may be able to create life in the laboratory under conditions thought to prevail on the early Earth—I predict we'll do this within fifty years—but that tells us only that it *could* happen, not how it *did* happen. Like historians lacking data on crucial events (was there a real Homer who wrote the *Iliad* and the *Odyssey*?), students of historical sciences like cosmology and evolutionary biology are often forced to live with uncertainty. (The un-certainty is not about *everything*, however: we know when both the universe and life on Earth began; we're just not sure how.) Living with uncertainty is hard for many people, and is one of the reasons why people prefer religious truths that are presented as absolute. But many scientists (I am one) share the feelings of Richard Feynman, who expressed his comfort with ignorance in an interview with the BBC:

> I can live with doubt, and uncertainty, and not knowing. I think it's much more interesting to live not knowing than to have answers which might be wrong. I have approximate answers and possible beliefs and different degrees of certainty about different things. But I'm not absolutely sure of anything, and there are many things I don't know anything about, such as whether it means anything to ask why we're here, and what the question might mean. I might think about it a little bit; if I can't figure it out, then I go on to something else. But I don't have to know an answer. I don't feel frightened by not knowing things, by being lost in the mysterious universe without having any purpose, which is the way it really is, as far as I can tell—possibly. It doesn't frighten me.

Feynman went a bit far in claiming that he wasn't absolutely sure of anything, for he surely knew that he'd die one day (sadly, that day came too soon), and that he'd fall if he stepped off his roof. But his statement does encapsulate the doubt that is endemic—and necessary—in science. It's not only endemic: it's one of science's attractions. A scientist lacking a big, juicy unsolved problem is a scientist bereft. H. L. Mencken compared the scientific investigator to "the dog sniffing tremendously at an infinite series of rat-holes," and that was meant as a compliment. Our living with doubt contrasts strongly with the way many people regard their religion. True, some believers wrestle with doubt and uncertainty, but it's a mind-set that's neither encouraged, common, nor comfortable. Clerics and coreligionists usually urge the doubter to wrestle with those uncertainties and, in the end, resolve them. But with religion, there's no real way to resolve them, for there's no procedure for checking whether your doubts are justified. You're then faced with either returning to your original faith or becoming an unbeliever.

Collectivity

One of the best parts of science's toolkit is its international character, or rather, its transcendence of nationality, for although there are scientists throughout the world, we all work by the same set of rules. The participants in the discovery of the Higgs boson, for instance, came from 110 countries, with 20 of those nations being official collaborators in the project. When I visit Turkey, Russia, Austria, or India, I can discuss my work with my colleagues without any cultural awkwardness or misunderstandings. Although scientists come in all faiths, including no faith at all, there is no Hindu science, no Muslim science, and no Jewish science. There is only science, combining brainpower from the whole world to produce one accepted body of knowledge. In contrast, there are thousands of religions, most differing profoundly in what they see as "true."

Curiously, the earliest scientific test I know of is actually described in the Bible. If you look at the First Book of Kings (18:21–40), you'll find a controlled experiment designed to reveal which god is real—Baal or the Hebrew god Yahweh. The test, proposed by the prophet Elijah, involved two bullocks, each killed, dressed, and placed on a separate pyre. Worshippers of each god were then told to ask their deity to ignite the pyre. Importuning Baal had no effect, even when his acolytes cut themselves with knives and lancets. But the Hebrew god came through, for even when his pyre was drenched with water, it burst into flame. Score one for Yahweh, scientifically shown to be the true god. In this case the penalty for being wrong was severe: the worshippers of Baal were summarily slain. But this story also invalidates the common claim that God won't be tested, for God willingly participated in this experiment.

Along with the methodology that I've described as "science" come the accoutrements of *professional* science: having grant support for one's research (usually from peer-reviewed applications to the government), submitting papers that are refereed by your peers before publication, having a job that can be categorized as "a scientist," and so on. But these are ancillary to the methods themselves, which are in fact used by many people who aren't usually considered scientists. In fact, I see science, conceived broadly, as *any* endeavor that tries to find the truth about nature using the tools of reason, observation, and experiment. Archaeologists use science when they date and study ancient civilizations. Linguists use science when they reconstruct the historical relationships between languages. Historians use science when they try to discover how many people died in the Holocaust, or refute the claims of Holocaust deniers. Art historians use science when dating

paintings or trying to discern whether one is a forgery. Economists and sociologists use science when they try to understand the causes of social phenomena, although "truths" in those areas can be elusive. Native peoples use
science when figuring out which local plants are useful in illness. (The use
of quinine to cure malaria, for instance, was derived from the Quechua of
Peru, who made an early version of "tonic" by mixing sweetened water with
the bitter ground bark of the cinchona tree.) Even biblical scholars use science when reconstructing how and when the Bible was written. Not all of
these areas, of course, are entirely scientific: much of the writing about history, for instance, involves untestable speculation about what caused various events.

The methods of science aren't even limited to academics. Car mechanics
use science when working out a problem in your electrical system, for they
make and test hypotheses about where the defect lies. Plumbers use science,
and their knowledge of hydraulics, when finding the source of leaks. The
kinship between "professional" science and plumbing was engagingly described by Stephen Jay Gould. In 1981, Gould was in Little Rock, Arkansas,
testifying in the famous trial of *McLean v. Arkansas,* during which a federal
judge adjudicated (and eventually rejected) a state law requiring "balanced
treatment" of evolution and creationism in public schools. On that visit
Gould encountered a plumber:

> As I prepared to leave Little Rock last December, I went to my hotel room
> to gather my belongings and found a man sitting backward on my com-
> mode, pulling it apart with a plumber's wrench. He explained to me that
> a leak in the room below had caused part of the ceiling to collapse and he
> was seeking the source of the water. My commode, located just above,
> was the obvious candidate, but his hypothesis had failed, for my equip-
> ment was working perfectly. The plumber then proceeded to give me a
> fascinating disquisition on how a professional traces the pathways of wa-
> ter through hotel pipes and walls. The account was perfectly logical and
> mechanistic: it can come only from here, here, or there, flow this way or
> that way, and end up there, there, or here. I then asked him what he
> thought of the trial across the street, and he confessed his staunch cre-
> ationism, including his firm belief in the miracle of Noah's flood.

As a professional, this man never doubted that water has a physical source and a mechanically constrained path of motion—and that he could use the principles of his trade to identify causes. It would be a poor (and unemployed) plumber indeed who suspected that the laws of engineering had been suspended whenever a puddle and cracked plaster bewildered him. Why should we approach the physical history of our earth any differently?

This anecdote shows not only the continuity of scientific methods (and "ways of knowing") across disparate areas, but also the disparity between science and religion embodied in a plumber who believed in Noah's flood.

What Is Religion?

Defining "religion" is a thankless task, for no single definition will satisfy everyone. Belief in a god would seem mandatory, but some groups that look like religions, such as Jainism, Taoism, Confucianism, and Unitarian Universalism, don't even have that. Other "religions," like Tibetan Buddhism, may not worship gods, but do accept supernatural phenomena like karma and reincarnation.

Rather than argue semantics, I'll choose a definition that fits most people's intuitive conceptions of religion, and certainly corresponds to the tenets of the three Abrahamic faiths—Judaism, Christianity, and Islam—that comprise about 54 percent of the world's inhabitants. This is also the form of religion that most often conflicts with science. The definition is taken from the *Oxford English Dictionary:*

> **Religion.** Action or conduct indicating belief in, obedience to, and reverence for a god, gods, or similar superhuman power; the performance of religious rites or observances.

One can derive three characteristics of religion from this definition, all part of the Abrahamic faiths. The first is *theism:* the claim that God interacts with the world. The notion of "superhuman power" implies that God's power is exercised, and the ideas of obedience and reverence, as well as performance of rites, imply that God must not only observe you but judge you, and his approval implicitly carries rewards or punishment. This means that I am considering religion as largely theistic, rather than a deistic belief in a remote, noninteractive God. As we'll see, few religionists are strict deists anyway. But even deism, though denying God's influence in the world, conflicts with science by making claims about God's existence, and often about his creation of the universe.

The second feature of religion is its embrace of a *moral system*. If the supernatural agent confers or denies approval based on obedience, that means there are behaviors and thoughts that are either worthy or unworthy of that approval, including obedience itself. This yields a framework of divinely based morality. Even faiths that lack gods, such as Taoism and Jainism, could still be considered philosophies with moral codes. (Jains, for instance, devoutly abjure harming any creatures, including insects, and even try to avoid injuring plants!)

Codes of morality imply the third trait of religion: the idea that God interacts directly with *you* in a personal relationship. In *The Varieties of Religious Experience*, William James saw the ideas of a moral code and a personal connection to God as the nucleus of all religions:

> [T]here is a certain uniform deliverance in which religions all appear to meet. It consists of two parts [an uneasiness and its solution]:
>
> 1. The uneasiness, reduced to its simplest terms, is a sense that there is something wrong about us as we naturally stand.
>
> 2. The solution is a sense that we are saved from the wrongness by making proper connection with the higher powers.

Finally, what do we mean by a "supernatural agent"? As we'll see, the term "supernatural" is slippery, for even supernatural powers can affect natural processes, bringing the supernatural into the realm of empirical study. I'll rely again on the *Oxford English Dictionary*'s definition of the adjective: "Belonging to a realm or system that transcends nature, as that of divine, magical, or ghostly beings; attributed to or thought to reveal some force beyond scientific understanding or the laws of nature; occult, paranormal." Here "beyond scientific understanding" means "outside the realm of the material world." As a supernatural being, God is often seen as a "bodiless mind," but one with humanlike emotions.

From now on I'll concentrate on religions that make empirical claims about the existence of a deity, the nature of that deity, and how it interacts with the world. But what do we mean by "claims"? Are they the claims of the church itself (that is, official doctrine and dogma), the claims of theologians (which, of course, differ, even among clerics within a faith), or the claims of regular believers, which needn't coincide with those of either theologians or church doctrine? We all know Catholics, for instance, who consider themselves members of the church although rejecting its doctrines on homosexuality and abortion, as well as the theory of evolution, which is accepted by the Vatican but rejected by many Catholics. When I discuss the claims of "religion," I'll simply go back and forth between theologians, believers, and dogma, trying to make clear which I'm discussing. Except for those rarefied theologians whose claims are either terminally obscure or close to atheism, it makes little difference, for believers, dogma, and theologians alike make existence claims and promote "ways of knowing" that make their faith incompatible with science. But do religions really make such claims? One needn't look far to discover that most do, although more sophisticated believers and theologians tend to downplay that fact.

Does Religion Look for Truth?

It seems obvious that if religion is based on the existence of a god, then that is a contention about reality, and such a reality constitutes a basis for belief. In other words, the existence of God is taken as a fact. Surprisingly, some theologians come close to denying this, saying that God cannot be described, or is beyond all ken, thus rejecting any empirical claims about a deity save its existence. Religion, they say, has little or nothing to do with facts, but is about morals, building a community, or finding a way of life. Here are two examples of such denial from believers, the first from Francis Spufford, a Christian, and the second from Reza Aslan, a Muslim:

Religion isn't a philosophical argument, just as it isn't a dodgy cosmology, or any other kind of alternative to science. In fact, it isn't primarily a
system of propositions about the world at all. Before it is anything else, it
is a structure of feelings, a house built of emotions. You don't have the
emotions because you've signed up to the proposition that God exists;
you entertain the proposition that God exists because you've had the
emotions.

It is a shame that this word, *myth*, which originally signified nothing
more than stories of the supernatural, has come to be regarded as synonymous with falsehood, when in fact myths are always true. By their very
nature myths inhere both legitimacy and credibility. Whatever truths
they convey have little to do with historical fact. To ask whether Moses
actually parted the Red Sea, or whether Jesus truly raised Lazarus from
the dead, or whether the word of God indeed poured through the lips of
Muhammad, is to ask totally irrelevant questions. The only question that
matters with regard to a religion and its mythology is "What do these
stories mean?"

But regardless of whether the emotions precede the belief (William
James's thesis in *The Varieties of Religious Experience*) or the belief yields
the emotions, the belief is still required to channel your emotions into a
moral code, a way of life, and religiously based actions. Spufford, after all,
has to "entertain the proposition" of God. Aslan reduces the Quran and the
Bible to collections of metaphors on which one can base a philosophy, but
hardly a religion. One can only imagine what most Muslims would say
about Aslan's contention, in a book about Islam, that it's irrelevant whether
Muhammad was really a prophet of God. Such a statement would get one
killed if uttered publicly in some Muslim lands.

It would seem unnecessary to document the importance of empirical
claims about God, except for those vociferous and liberal theologians who
argue that faith doesn't depend on statements about reality. Let's begin with
the Bible, which clearly grounds Christianity on the Resurrection, a supposedly historical event that has become the linchpin of virtually all Christian

faith: "Now if Christ be preached that he rose from the dead, how say some among you that there is no resurrection of the dead? But if there be no resurrection of the dead, then is Christ not risen: and if Christ be not risen, then *is* our preaching vain, and your faith *is* also vain."

And this is echoed by modern religious scholars, including Richard Swinburne, one of the world's most respected philosophers of religion:

> For the practices of the Christian religion (and of any other theistic religion) only have a point if there is a God—there is no point in worshipping a non-existent creator or asking him to do something on earth or take us to heaven if he does not exist; or trying to live our lives in accord with his will, if he has no will. If someone is trying to be rational in practicing the Christian (or Islamic or Jewish) religion, she needs to believe (to some degree) the creedal claims that underlie the practice.

Mikael Stenmark, a professor of philosophy and religion at Uppsala University in Sweden, and dean of its Faculty of Theology, is even more explicit. (His book *Rationality in Science, Religion, and Everyday Life* was awarded a Templeton Prize in 1996 for "outstanding books in theology and the natural sciences.")

> A religion therefore contains also (d) *beliefs about the constitution of reality.* . . . According to the Christian faith, our problem is that although we have been created in the image of God we have sinned against God and the cure is that God, through Jesus Christ, provides forgiveness and restoration. But for this cure to work it appears that at least it must be true that God exists, that Jesus Christ is the son of God, that we are created in the image of God, that God is a creator, that God wants to forgive us, and that God loves us. Hence it seems as if Christianity, and not only science, has an *epistemic* goal, that is, it attempts to say something true about reality. If so, a religious practice like Christianity is meant to tell us something true about who God is, what God's intentions are, what God has done, what God values, and how we fit in when it comes to these intentions, actions, and values.

John Polkinghorne, an English physicist who left Cambridge to become an Anglican priest, later became president of Queens College, Cambridge, and wrote several dozen books on the relationship of science and religion. Polkinghorne too won a Templeton Prize, and emphasized the need for an empirical grounding of faith:

> The question of truth is as central to [religion's] concern as it is in science. Religious belief can guide one in life or strengthen one at the approach of death, but unless it is actually true it can do neither of these things and so would amount to no more than an illusionary exercise in comforting fantasy.

Ian Barbour, who died in 2013, was an American professor of religion (and another winner of the Templeton Prize) who specialized in the relationship between science and religion:

> A religious tradition is indeed a way of life and not a set of abstract ideas. But a way of life presupposes beliefs about the nature of reality and cannot be sustained if those beliefs are no longer credible.

Finally, we have a joint statement by Francis Collins, a born-again Christian who directs the U.S. National Institutes of Health, and Karl Giberson, a Christian physicist. They were respectively once president and vice president of the accommodationist organization BioLogos:

> Likewise, religion in almost all of its manifestations is more than just a collection of value judgments and moral directives. Religion often makes claims about "the way things are."

Existence Claims: Is There a God?

Some faith claims are more important than others, but nearly all theists have at least one or two bedrock beliefs that support their religion. The most important, of course, include the existence of a god, whether there is only one of them or, in polytheistic faiths like Hinduism, a panoply of gods with different abilities. Existence claims about gods are clearly empirical claims—claims that require some kind of evidence—and although they can be hard to test depending on the kind of god you worship, advocates of theism argue that God's interventions in the universe should be detectable. At the very least, those theists should be able to describe what the world would be like had it arisen in a purely naturalistic manner, and if their god didn't exist.

Many surveys show that belief in gods is universal and strong. A 2011 survey of belief in twenty-two countries, for instance, found that 45 percent of all people asked agreed with the statement "I definitely believe in a God or a Supreme Being." But there was wide variation among nations, ranging from 93 percent agreement in Indonesia to only 4 percent in Japan. (Besides Turkey and Indonesia, "Muslim countries" weren't surveyed, nor were any in Africa, though belief in both regions is surely very high.) European countries were on the low side, with between 20 percent and 30 percent of people being "definite" believers, with Great Britain coming in at 25 percent. The United States, the most religious of First World countries, ranked seventh, with 70 percent espousing definite belief in God. (Definite *nonbelief,* by the way, was expressed by 18 percent of Americans—about half the level found in France, Sweden, and Belgium.)

God, of course, can be construed along a continuum from the traditional bearded man in the sky to the ineffable "ground of being" of modern theologians. But three surveys conducted by the International Social Survey Program between 1991 and 2008 narrowed this down, asking people in thirty countries whether they believed in a *personal god* "who concerns himself with every human being personally." This goes further than just assuming that God affects the world. But the results resembled those given above: there was wide variation, ranging from 20 to 30 percent in most European countries to 68 percent in the United States, but there was also widespread acceptance, in studies spanning two decades, of an involved and intervening God. Clearly most of those who accept God are theists, not deists.

Right before I wrote these paragraphs, a member of the Jehovah's Witnesses e-mailed me an article, "The Untold Story of Creation," which was quite specific about God's nature:

God is a person, an individual. He is not a vague force devoid of personality, floating aimlessly in the universe. He has thoughts, feelings, and goals.

The more intellectual believers would sneer at such a description, claiming that God is not at all like a person, and that their own nebulous and impersonal deity is the "correct" description of God. (How they know this is never specified.) But that's not the take of the many highly respected and nonliteralist theologians who still imbue God with personlike qualities. The list of God's attributes from *The National Catholic Almanac* reads like a dictionary definition:

> Attributes of God. Though God is one and simple, we form a better idea by applying characteristics to Him, such as: almighty, eternal, holy, immortal, immense, immutable, incomprehensible, ineffable, infinite, intelligent, invisible, just, loving, merciful, most high, most wise, omnipotent, omniscient, omnipresent, patient, perfect, provident, self-dependent, supreme, true.

Here's how Richard Swinburne sees God:

> I take the proposition "God exists" (and the equivalent proposition "There is a God") to be logically equivalent to "there exists necessarily a person without a body (i.e. a spirit) who necessarily is eternal, perfectly free, omnipotent, omniscient, perfectly good, and the creator of all things." I use "God" as the name of the person picked out by this description.

Alvin Plantinga is America's equivalent to Swinburne: a respected philosopher/theologian, once regional president of the American Philosophical Association. What does he say about God?

> What [Daniel Dennett] calls an "anthropomorphic" God, furthermore, is precisely what traditional Christians believe in—a god who is a *person*, the sort of being who is capable of knowledge, who has aims and ends,

and who can and in fact does act on what he knows in such a way as to try to accomplish those aims.

For every theological claim that God is a spirit or force about whom we can say little—except that he exists—I can adduce several statements from theologians and believers swearing that God resembles a powerful but bodiless person, with human emotions, motivations, and a loving personality. This view of the deity is not so different from that of the Jehovah's Witnesses mentioned above, or even the one described to young children in *Bruce and Stan's Pocket Guide to Talking with God:*

> It's really important to understand that God is not an impersonal force. Even though He is invisible, God is personal and He has all the characteristics of a person. He *knows,* he *hears,* he *feels* and he *speaks.*

Liberal theologians like Karen Armstrong and David Bentley Hart, who maintain that God is not like this at all, either dismiss the universal belief in a personal God or claim, on no convincing grounds, that it's wrong. And even if you think that the nebulous "ground of being" God is the most convincing God, you're ignoring the beliefs of those who actually inject their dogma into the public arena. Certainly one can deal with the "best" arguments for God—which invariably turn out to be the ones so fuzzy that they're the least capable of being falsified, much less understood—but it's more important to deal with religious beliefs as they're held by the vast majority of people on Earth.

Other Empirical Claims of Religion

Beyond God, what are the other truths that religions hold dear? I've taken one version of the Nicene Creed—a statement recited weekly in many Christian churches, and one of many such creeds maintained by different religions—and simply put in bold its truth claims. While many Christians may piously mouth these words without believing them, many believers certainly see them as true. And virtually every word in this creed is a claim about the universe:

I believe in one God,
the Father almighty,
maker of heaven and earth,
of all things visible and invisible.
I believe in one Lord Jesus Christ,
the Only Begotten Son of God,
born of the Father before all ages.
God from God, Light from Light,
true God from true God,
begotten, not made, consubstantial with the Father;
through him all things were made.
For us men and for our salvation
he came down from heaven,
and by the Holy Spirit was incarnate of the Virgin Mary,
and became man.
For our sake he was crucified under Pontius Pilate,
he suffered death and was buried,
and rose again on the third day
in accordance with the Scriptures.
He ascended into heaven
and is seated at the right hand of the Father.
He will come again in glory
to judge the living and the dead
and his kingdom will have no end.
I believe in the Holy Spirit, the Lord, the giver of life,
who proceeds from the Father and the Son,
who with the Father and the Son is adored and glorified,
who has spoken through the prophets.
I believe in one, holy, catholic and apostolic Church.
I confess one Baptism for the forgiveness of sins
and I look forward to the resurrection of the dead
and the life of the world to come. Amen.

In summary, the Creed claims a monotheistic God, who nevertheless somehow consists of three parts (including Jesus and the Holy Spirit); the creation of the universe by that God; and the sending of his son—born of a virgin—as an earthly sacrifice to redeem believers from sin. It further asserts that God's son (who was also God) was crucified but resurrected on the third day after death and, although presently residing in heaven, will one day return, raising the dead and pronouncing on all an eternal sentence of either bliss or fire. Baptism is deemed essential for entry to heaven. These are all empirical statements about reality: they are either true or false, even if some are hard to investigate.

These claims, of course, absolutely conflict with those of other faiths. Jews, Muslims, Hindus, and Sikhs don't recognize Jesus as the Messiah. Muslims believe that those who do so will spend eternity in hell. Doesn't choosing among such faiths require a way to evaluate whether this dogma is true?

Still, how do we know whether those who recite things like the Nicene Creed take its words literally? Look at the polls, many of which show that such literalism is widespread. The most recent survey of Americans, a Harris poll of representative citizens taken in 2013, shows a surprisingly large number of people who accept supernatural claims. Besides the 54 percent who are "absolutely certain there is a God" (an additional 14 percent are "somewhat certain"), beliefs in things like the divinity of Jesus, miracles, the existence of heaven, hell, Satan, angels, and the survival of the soul after death are all above 56 percent. In contrast, only 47 percent believe in Darwin's theory of evolution (we scientists prefer to use "accept" rather than "believe" when we speak of scientific theories). Further, 39 percent of Americans conceive of God as male, but only 1 percent as female (38 percent see God as "neither"), supporting the idea that if people see God as a bodiless person, it often has genitalia. As for the veracity of scripture, 33 percent accept the Old Testament as being "completely the Word of God," while 31 percent gave the same answer for the New Testament. Remember, these statistics are from a sample of *all* Americans, not just believers. Scriptural literalism is certainly widespread in the United States—in fact, depending on the claim, it's often a majority view.

Readers from the United Kingdom are undoubtedly thinking, "Well, that's the hyperreligious United States. We're not *nearly* that pious." And that's true, but Britain still shows a surprisingly high level of religious literalism. In 2011,

Julian Baggini, an atheist philosopher who was nevertheless sympathetic to religion, grew tired of the claims of "strident" atheists, who, he said, wrongly saw Christianity as depending heavily on facts. To get data on the content of religious belief, Baggini surveyed nearly eight hundred churchgoing Christians in an online poll in the *Guardian*. Now, this is hardly the kind of rigorous "scientific" poll conducted by Harris or Gallup, and the results could have been biased by a greater willingness of more religious people to respond. Nevertheless, Baggini was astonished at the literalism of those who answered. Asked why they went to church, for instance, 66 percent responded that they did so "to worship God," while only 20 percent went for the "feeling of community" (so much for claims that the social aspects of religion far outweigh its dogma!). There was also widespread agreement that the stories in Genesis, such as Adam and Eve, really happened (29 percent), that Jesus performed miracles such as that of the loaves and fishes (76 percent), that Jesus's death on the cross was necessary for forgiveness of human sin (75 percent), that Jesus was bodily resurrected (81 percent), and that eternal life required accepting Jesus as lord and savior (44 percent). Chastened, Baggini retracted his previous views:

> So what is the headline finding? It is that whatever some might say about religion being more about practice than belief, more praxis than dogma, more about the moral insight of mythos than the factual claims of logos, the vast majority of churchgoing Christians appear to believe orthodox doctrine at pretty much face value. . . . This is, I think, a firm riposte to those who dismiss atheists, especially the "new" variety, as being fixated on the literal beliefs associated with religion rather than ethos or practice. It suggests that they are not attacking straw men when they criticise religion for promoting superstitious and supernatural beliefs.

Data are sparser from the rest of the world, but also show a high degree of religious literalism, especially outside Europe. The 2011 Ipsos/Reuters poll showed that belief in the existence of heaven or hell was held by 19 percent of the combined inhabitants of twenty-three countries surveyed, ranging from only 3 percent of Swedes to 62 percent of Indonesians. The same belief was entertained by 41 percent of Americans and only 10 percent of

Britons. We see some disparities between these results and those of the Harris poll, which showed a higher percentage of Americans believing in heaven and hell. These disparities might be due to the way the questions are asked, and should make us wary of taking any statistic as a precise estimate. Nevertheless, no polls show that most believers see scripture metaphorically rather than literally.

The world's Muslims are especially pious and literalistic. It's no surprise that a 2012 Pew survey of thirty-eight thousand professed Muslims in thirty-eight countries showed that belief in God and in Muhammad as his prophet was nearly universal (the median percentage ranged from 85 percent in southeastern Europe to between 98 percent and 100 percent in the Middle East and North Africa). But it might be a surprise to those unfamiliar with Islam that in all countries surveyed, more than half of Muslims asserted that the Quran "should be read literally, word for word": figures ranged from 54 percent in the Democratic Republic of the Congo to 93 percent in Cameroon (data were not available for the Middle East). Muslim belief in angels ranged from a low of 42 percent in Albania to a high of 99 percent in Afghanistan (90 percent in the United States), with twenty-three of thirty-eight countries showing more than 80 percent.

The survey shows that on the whole, most Muslims are Quranic literalists, even more faithful to scripture than are highly religious Americans. Islamic literalism is one reason why, when Muslims perceive an offense to their faith, like the Danish cartoons that mocked Muhammad, they rise up en masse, often in violent sprees. One must take seriously the claim that *they really believe what they say they believe,* and that faith, not reason, can be a major cause of religious malfeasance. Islam is, of course, not unique in this way; as we'll see in the final chapter, the dangers of faith are inherent in many other religions as well.

It is a staple of accommodationists, and of those atheists who "believe in belief," to exculpate religion by ascribing what are clearly religiously motivated acts to "politics" or "social dysfunction." (In many Muslim countries, however, there's virtually no demarcation between religion, politics, and social mores.) This is simply an extension of the claim that religion doesn't really involve truth claims about the universe. In a debate with Steven Pinker about "scientism"—the notion that science often intrudes into areas where it doesn't belong—the *New Republic* editor Leon Wieseltier wrote, "Only a small minority of believers in any of the scriptural religions, for example, have ever taken scripture literally." But that's simply wrong. Perhaps some Christians see the Bible largely as allegory, but there are some nonnegotiable beliefs that are virtually diagnostic of each religion. William Dembski, a Southern Baptist and prominent advocate of intelligent design creationism, has specified the "non-negotiables of Christianity" as these: divine creation, reflection of God's glory in the world, the exceptionalism of humans made in the image of God, and the Resurrection of Christ. These constitute the epistemic claims of faith, and virtually every believer entertains some. (For Christians, the ultimate redoubt is often the Resurrection.) As I often say, some believers are literalists about nearly everything, but nearly every believer is a literalist about something.

Is Scripture Literal or Allegorical?

This brings us to the thorny question of metaphor and allegory (allegory is just extended metaphor: an entire story that is not meant to be taken literally, but symbolizes an underlying message). A recurrent pattern in theology is this: as branches of science—evolutionary biology, geology, history, and archaeology—have disproved scriptural claims one by one, those claims have morphed from literal truths into allegories. This is one of the big differences between science and religion. When a scientific claim is disproved, it goes into the dustbin of good ideas that simply didn't pan out. When a religious claim is disproved, it often turns into a metaphor that imparts a made-up "lesson." Although some biblical events are hard to see as allegories (Jonah's ingestion by a fish and Job's trials are two of these), the theological mind is endlessly creative, always able to find a moral or philosophical point in fictitious stories. Hell, for instance, has become a metaphor for "separation from God," and now that we know that Adam and Eve cannot have been the literal ancestors of all living people (see chapter 3), the "original sin" they bequeathed is seen by some believers as a metaphor for our evolved selfish nature.

Further, many liberal believers are affronted by claims that nearly anything in the Bible should be taken literally. One of their most common arguments against such literalism is this: "The Bible is not a textbook of science." When I see that phrase, I automatically translate it as, "The Bible is not entirely true," for that is what it means. The "nontextbook" claim, of course, is a rationale for believers to pick and choose what they consider *really* true in scripture—or, for liberal Muslims like Reza Aslan, in the Quran.

Indeed, even saying that there's a historical *tradition* of taking scripture literally can deeply upset "modern" believers, for the fashion is to argue that literalism is purely a modern phenomenon. When I wrote on my Web site that the story of Adam and Eve could not be literally true, for evolutionary genetics had shown that the population of modern humans was always much larger than two, the writer Andrew Sullivan took me to task for even suggesting that believers saw the First Couple as historical figures:

> There's no evidence that the Garden of Eden was always regarded as figurative? Really? Has Coyne read the fucking thing? I defy anyone with a brain (or who hasn't had his brain turned off by fundamentalism) to think it's meant literally.

Yet for centuries, Christians, and that includes the Catholic Church, to which Sullivan belongs, took the story of Adam and Eve as the sole ancestors of humanity literally. And no wonder, for the description in the Bible is straightforward, without the slightest hint that it's an allegory.

Now, when Jesus recites parables, like that of the Good Samaritan, it's clear that he's simply telling a story to make a point. But that's not the way that Genesis reads. Catholics have in fact always adhered literally to *religious monogenism*, the biological descent of all humans from Adam and Eve. The reality of the Garden of Eden, the Fall, and Adam and Eve as our ancestors was accepted by early theologians and church fathers like Augustine, Aquinas, and Tertullian, although some, like Origen, were unclear on the issue. In 1950, however, Pope Pius XII affirmed monogenism in his encyclical *Humani Generis*. After asserting that the church didn't oppose research and discussion of evolution—so long as everyone agreed that, during the process, only humans were given a soul by God—the pope denied such latitude about Adam and Eve:

When, however, there is question of another conjectural opinion, namely polygenism [our descent from ancestors beyond Adam and Eve], the children of the Church by no means enjoy such liberty. For the faithful cannot embrace that opinion which maintains that either after Adam there existed on this earth true men who did not take their origin through natural generation from him as from the first parent of all, or that Adam represents a certain number of first parents. Now it is in no way apparent how such an opinion can be reconciled with that which the sources of revealed truth and the documents of the Teaching Authority of the Church propose with regard to original sin, which proceeds from a sin actually committed by an individual Adam and which, through generation, is passed on to all and is in everyone as his own.

There is no room for waffling here. The authority of the church insists that a historical Adam committed a sin passed on to his offspring—as if sin were a gene that never gets lost—and those sinful offspring grew into all of humanity.

The historical emphasis on the existence of a literal Adam and Eve, and the couple's crucial position in theology, is emphasized by the historian David Livingstone in his book *Adam's Ancestors:*

Regardless of how differently the Garden of Eden may have been conceived from ancient times through the medieval period to more recent days, and no matter the differences in computations of the creation date of the earth, the idea that every member of the human race is descended from the biblical Adam has been a standard doctrine in Islamic, Jewish and Christian thought. In this respect, if in no other, the catechisms of the seventeenth-century Westminster divines can be taken to speak for them all when they declare that "all mankind" descended from Adam "by ordinary generation." People's sense of themselves, their understanding of their place in the divinely ordered scheme of things, their very identity as human beings created in the image of God, thus rested on a conception of human origins that assumed the literal truth of the biblical narrative and traced the varieties of the human race proximately to the three sons of Noah and ultimately to Adam and Eve.

I dwell on Adam and Eve for two reasons. The first is simply to show that despite the claims of religious liberals like Sullivan, there's no denying that over history much of the Bible has been seen literally, particularly when—as in the case of the First Couple—an important doctrine is at stake. I often hear theologians argue that their predecessors like Aquinas and Augustine were not literalists, and that literalism began only in the nineteenth or twentieth century. But that's a distortion of history, one designed to save churches from the embarrassment of having taken seriously stories now seen as palpably fictitious.

Saint Thomas Aquinas, for instance, is often praised for having argued that scripture can be read metaphorically. Such a claim, though, is inaccurate, easily dispelled if you simply read his writings. Aquinas actually argued that scripture could be read *both literally and metaphorically*. In other words, he waffled, but, importantly, emphasized that if there was a conflict between metaphorical and literal interpretations of the Bible, literalism must win.

Here, for example, is Aquinas discussing the reality of paradise, the abode of Adam and Eve, in *Summa Theologica*. Responding to the words of his predecessor Saint Augustine, Aquinas shows how historical truth trumps metaphor (my emphasis):

Augustine says (Gen. ad lit. viii, 1): "Three general opinions prevail about paradise. Some understand a place merely corporeal; others a place entirely spiritual; while others, whose opinion, I confess, pleases me, hold that paradise was both corporeal and spiritual."

I answer that, As Augustine says (De Civ. Dei xiii, 21): "Nothing prevents us from holding, within proper limits, a spiritual paradise; so long as we believe in the truth of the events narrated as having there occurred." For whatever Scripture tells us about paradise is set down as matter of history; and *wherever Scripture makes use of this method, we must hold to the historical truth of the narrative as a foundation of whatever spiritual explanation we may offer.*

Aquinas believed not only in paradise, but also in the instantaneous creation of species and of Adam and Eve as humanity's ancestors, as well as in a young Earth (less than six thousand years old) and the literal existence of

Noah and his great flood. Further, Aquinas was obsessed with angels. Not only did he see them as real but devoted a large section of the *Summa Theologica* ("Treatise on the Angels") to their existence, number, nature, how they move, what they know, and what they want. The philosopher Andrew Bernstein describes such theological analysis of arcane and unevidenced claims as "the tragedy of theology in its distilled essence: The employment of high-powered human intellect, of genius, of profoundly rigorous logical deduction—studying nothing."

Saint Augustine of Hippo, who commented extensively on Genesis, was quite explicit that the text, though it had a spiritual message, was based on historical events:

> The narrative indeed in these books is not cast in the figurative kind of language you find in the Song of Songs, but quite simply tells of things that happened, as in the books of the Kingdoms and others like them. But there are things being said with which ordinary human life has made us quite familiar, and so it is not difficult, indeed, it is the obvious thing to do, to take them first in the literal sense, and then chisel out from them what future realities the actual events described may figuratively stand for.

Augustine was also a literalist about many things later refuted by science: a young Earth, instantaneous creation, the historical reality of Adam and Eve, paradise, and Noah and his Ark. It's ironic that both he and Aquinas are constantly touted by accommodationists as having a "nonliteral" theology that is completely compatible with science in general and with evolution in particular. Such a claim can be made only by those who haven't read these theologians or are dedicated to whitewashing church history.

I could go on, but two more examples will suffice. The Protestant reformer John Calvin believed in the virginity of Mary, a historical Adam and Eve, and a literal hell. Like Aquinas, he also believed that heretics should be killed. As for metaphorical interpretation of the Quran, that's simply not on the menu: as we saw above, the majority of the world's Muslims see that document as literally true.

Sullivan's rage about Adam and Eve raises my second point. If you want to read much of the Bible as allegory, you must overturn the history of the-ology, rewriting it to conform to your liberal, science-friendly faith. Besides pretending that you're following in the tradition of ancient theologians, you must also explain the way you can discern truth amid the metaphors. What is allegory and what is real? How do you tell the difference? This is particu-larly difficult for Christians, because the historical evidence for Jesus—that is, for a real person around whom the myth accreted—is thin. And evidence for Jesus as the son of God is unconvincing, resting solely on the assertions of the Bible and interpretations of people writing decades after the events described in the Gospels.

If faith is often grounded on facts, we might expect one of two results if those facts were shown to be wrong: either people would abandon their faith—or some parts of it—or they would simply deny the evidence that contradicted their beliefs.

There isn't much data on the first possibility, but there's some suggestion that at least major parts of faith are resistant to scientific disproof. As we've seen, 64 percent of Americans would retain a religious belief even if science disproved it, while only 23 percent would consider changing that belief. The results were only slightly less disheartening in Julian Baggini's online sur-vey of British churchgoing Christians, 41 percent of whom either agreed or tended to agree with the statement "If science contradicts the Bible, I will believe the Bible, not science."

Evolution: The Biggest Problem

The clearest example of religion's resistance to science is, of course, its atti-tude toward evolution. While not the only scientific theory that contradicts scripture, evolution has implications, involving materialism, human excep-tionalism, and morality, that are distressing to many believers. And yet it is supported by mountains of scientific data—at least as much data as support the uncontroversial "germ theory" that infectious diseases are caused by microorganisms.

And indeed, evolution is largely rejected by the faithful. Among twenty-three countries surveyed in a 2011 Ipsos/Reuters poll on acceptance of hu-man evolution, 28 percent of all people rejected it in favor of creationism, with the rejection higher in more religious countries. Saudi Arabia and Turkey were the biggest deniers of evolution. (This relationship also holds among states within the United States: the most religious states show the most denial of evolution.) The situation is especially dire in the United States, a country considered scientifically advanced. Yet when it comes to evolution, many Americans remain in the Bronze Age. A 2014 Gallup poll of American attitudes toward human evolution showed that fully 42 percent were straight biblical young-Earth creationists, agreeing that humans were created in our present form within the last ten thousand years. Another 31 percent were "theistic evolutionists," accepting evolution with the caveat that it was supernaturally guided or prompted by God. And only 19 percent of Americans—fewer than one in five—accepted evolution the way biologists do, as a naturalistic, unguided process. These figures have remained almost constant over the last three decades, with perhaps a slight and recent rise in those accepting naturalistic evolution.

This rejection of evolution can't be explained simply by Americans' ignorance of the evidence. We live in an age of unprecedented science popularization: think of people like Richard Dawkins, Carl Sagan, David Attenborough, Neil deGrasse Tyson, and Edward O. Wilson. The evidence for evolution is everywhere—only a few clicks away on the Internet, a perusal of *National Geographic,* or a single push on the remote control. Yet evolution is rejected by Americans as strongly as it was three decades ago.

The reason is clear. When asked in 2007 why they denied evolution, Americans gave as the main reasons their belief in Jesus (19 percent of respondents), God (16 percent), or religion in general (16 percent), all exceeding those who thought there was "not enough scientific evidence" for evolution (14 percent). Other studies show that although religious people in the United States know slightly less about science than do nonbelievers, their knowledge about what the theory of evolution actually *says* is about the same. Nevertheless, regardless of how science-savvy they are, the religious deny the fact of evolution much more often than do the nonreligious; in fact, the more the faithful know about science, the more they reject evolution! This arrant rejection of facts is clearly based not on lack of education or ignorance, but on religious belief.

Indeed, faith can trump facts even when church authority accepts the facts. The Catholic Church, for instance, accepts a form of theistic evolution, largely naturalistic but still tweaked by God, who instilled souls in *Homo sapiens* at some point in our evolution. Nevertheless, 27 percent of American Catholics cling to biblical creationism, believing that humans were created instantaneously by God and have remained unchanged ever since. Resistance to evolution in America, then, can be laid completely at the door of religion. You can find some religions without creationism, but you can't find creationism without religion.

It's a useful exercise to ask religious people what it would take for them to either abandon the "nonnegotiables" of their faith—like the view that Jesus was divine or that the Quran is the word of Allah—or to give up their faith entirely. Very often you will get the answer "Nothing could make me give up those beliefs." As we'll see below, that's one of the many incompatibilities between the attitudes toward religious "truth" and scientific truth. Scientists are not only constantly looking for evidence that would prove their pet theories wrong, but often know exactly what kind of evidence would do it. There are no "nonnegotiables" in science.

Can You Have Faith Without Truth Claims?

Religion, of course, is not *solely* concerned with truth claims. As Francis Spufford noted, many—perhaps most—people aren't religious because they're convinced by their church's arguments for God and scripture. Often religion really *is* "a structure of feelings, a house built of emotions." Belief in God often comes not from evidence, but from teaching or indoctrination by peers, or some revelation that *seems* real. The "evidence," often confected by theologians who specialize in justifying beliefs acquired in childhood, comes after. Or perhaps never, for how many religious people are even acquainted with the arguments for God's existence, or with the particulars of their belief? A survey of Americans in 2010 found, for instance, that Christians were abysmally ignorant about the details and doctrines of Christianity: only 42 percent of Catholics could name Genesis as the first book of the Bible, while only 55 percent knew that the bread and wine of the Communion *become*, rather than symbolize, the body and blood of Christ.

Further, it's often argued that the social and emotional aspects of belonging to a faith, rather than its dogma, are the real motivating force for membership. The psychologist Jonathan Haidt, for example, sees religious communality as the main motivation for faith-based social action. This idea needs further exploration, particularly because some data contradict it. In Baggini's study of British Christians, for instance, the percentage of people who went to church to worship God far exceeded those who went to feel part of a community or receive spiritual guidance from readings and sermons.

But even if religion provides solace and social benefits, we need to know how much those benefits rest on the belief that your religion's claims are true. How many Christians would *remain* Christian were they to know for sure that Christ was neither divine nor resurrected but, as some biblical scholars like Bart Ehrman believe, simply an apocalyptic preacher of the ancient Middle East? How many Mormons would retain their faith were they to know for certain that Joseph Smith inscribed the golden plates supposedly presented by the angel Moroni? It's hard to answer such questions, but what we do know is this: many of those who abandoned their religion ascribe it not to losing their feeling of community, but to losing their belief in its doctrines.

In her acclaimed book *When God Talks Back*, the Stanford anthropologist Tanya Luhrmann concluded, "People come to faith not just because they decide that the propositions are true but because they experience God directly. They feel God's presence. They hear God's voice. Their hearts flood with an incandescent joy." Her thesis is that it takes hard work to learn how to converse with God. But would those who succeed experience that joy if they didn't think that they were actually talking to someone who *listened*? Surely not much bliss accrues to those who think they're talking only to themselves.

It appears that theologians are ambivalent toward the empirical claims of religion. When writing for academics or liberal clerics, they downplay those claims, but when talking to "regular" believers, they affirm that faith rests on assertions about what's real in the universe. Alvin Plantinga, for instance, argues in one book that the literal truth of the Bible is subordinate to its moral lessons:

> The aim is to discover what God is teaching us in a given passage, and to
> do so in the light of these assumptions; the aim is not to determine

whether what is taught is true, or plausible, or well supported by the arguments.

But only a year earlier, Plantinga claimed not only that God exists, but also that he has definite humanlike qualities:

> [In Christianity, Judaism, and Islam], theism is the belief that there is an all-powerful, all-knowing perfectly good immaterial person who has created the world, has created human beings "in his own image," and to whom we owe worship, obedience and allegiance. . . . Now God, according to theistic belief, is a person: a being who has knowledge, affection (likes and dislikes), and executive will, and who can act on his beliefs in order to achieve his ends.

The question to ask believers is this: "Does it really matter whether what you believe about God is true—or don't you care?" If it does matter, then you must justify your beliefs; if it doesn't, then you must justify belief itself.

As we'll soon learn, both theologians and garden-variety believers show strong resistance to arguments that strive to falsify the ideas and character of God, and often devise ways to justify religious claims in the face of counterevidence. This kind of defense suggests that people really *do* care that their religious beliefs are true and are not just psychologically useful fictions.

The Incompatibility

The next definition we need, of course, is one for "incompatibility," as there's always a way to construe that term that would make faith and fact seem compatible.

Let me first say what I *don't* mean by incompatibility. I don't mean *logical* incompatibility: that the existence of religion is simply and a priori incompatible with the practice of science. That's clearly wrong, for in principle there could be both science and a god to be worshipped. Nor do I mean *practical* incompatibility: the idea that one simply can't be a religious scientist or a science-friendly believer. That's clearly false as well, for there are many examples of both. Finally, I am not claiming that religious people are in general opposed to either science in general or the facts it reveals. Although some believers have problems with evolution and cosmology, the vast majority of religious people have no problem with issues like how genetics works, what causes disease and how to treat it, how molecules react chemically with one another, and the principles of aerodynamics. Indeed, nearly everyone in modern societies puts their trust in science every day.

My definition of "compatibility" is the second one given by the *Oxford English Dictionary* (the first is "participating in suffering, sympathetic"):

> Mutually tolerant; capable of being admitted together, or of existing together in the same subject; accordant, consistent, congruous, agreeable.

While religion and science could be considered "mutually tolerant," in that some scientists and believers tolerate each other's existence, and could even be seen as "capable of being admitted together," as with religious scientists, I don't see them as "existing together in the same subject" or as "accordant, consistent, congruous, agreeable."

My claim is this: science and religion are incompatible because they have different methods for getting knowledge about reality, have different ways of assessing the reliability of that knowledge, and, in the end, arrive at conflicting conclusions about the universe. "Knowledge" acquired by religion is at odds not only with scientific knowledge, but also with knowledge professed by other religions. In the end, religion's methods, unlike those of science, are useless for understanding reality. This form of incompatibility is the one expressed—albeit more humorously—by the science writer Natalie Angier in the quote that heads this chapter.

One might argue, using the dictionary definition, that religion and science are actually compatible because they are "consistent" in one respect: science's realm involves facts about the universe, while that of religion is supposedly limited to morals, meaning, and purpose. In other words, they are compatible because they are complementary. I will argue in the next chapter that this idea, made famous by Stephen Jay Gould, fails on two counts: religion *also* deals with facts about the universe, and even if religion claims to deal with the "big questions" about human purpose and value, so do other fields, like secular philosophy, that don't use the concept of a god.

Clearly, religions that don't make existence claims, like Taoism, Confucianism, and pantheism, aren't incompatible in the way I describe. But theistic faiths, those that posit God's intervention in the world, conflict with science on three levels: methodology, outcomes, and philosophy.

Looking at methodology, I claim that the difference between science and religion can be summarized in how their adherents answer the question "How would I know if I was wrong?"

The difference in methods yields a difference in outcomes. Because the ways that science and religion come to understand reality are at odds, they are expected to produce different outcomes: different "facts." To the extent that scientific facts contravene religious doctrines, this creates incompatibilities.

Finally, the first two incompatibilities lead to the third: a disparity in philosophy. Science has learned through experience that assuming the existence of gods and divine intervention has been of no value in helping us understand the universe. This has led to the working assumption—some might call it a "philosophy"—that supernatural beings can be *provisionally seen as nonexistent.* I'll take up these issues in order.

Conflicts of Method

The different methods that science and religion use to ascertain their "truths" couldn't be clearer. Science comprises an exquisitely refined set of tools designed to find out what is real and to prevent confirmation bias. Science prizes doubt and iconoclasm, rejects absolute authority, and relies on testing one's ideas with experiments and observations of nature. Its sine qua non is evidence—evidence that can be inspected and adjudicated by any trained and rational observer. And it depends largely on falsification. Nearly every scientific truth comes with an implicit rider: "Evidence X would show this to be wrong."

Religion begins with beliefs based not on observation, but on revelation, authority (often that of scripture), and dogma. Most people acquire their faith when young via indoctrination by parents, teachers, or peers, so that religious "truths" depend heavily on who spawned you and where you grew up. Beliefs instilled in this way are then undergirded with defenses that make them resistant to falsification. While some religious people do struggle with their beliefs, doubt is not an inherent part of belief, nor is it especially prized. No honors accrue to the Southern Baptist who points out that while there is plenty of evidence for evolution, there is none for the creation story of Genesis.

Some religious claims are untestable because they involve knowing about the irrecoverable past. There is almost no way to show, for instance, that Jesus was the son of God, that Allah dictated the Quran to Muhammad, or that the souls of Buddhists are reincarnated in other humans or animals. (There could, however, be at least some evidence for such claims, such as concordant eyewitness accounts of the miracles that supposedly accompanied Jesus's Crucifixion, including the darkness at noon, the rending of the Temple's curtain, the earthquakes, and the rising of saints from their graves. Unfortunately, the many historians of the time have failed to report these phenomena.) What science *can* do is point out the absence of evidence for such claims, taking them off the table until some hint of evidence arrives. When scientists don't know something, like the nature of the mysterious "dark matter" that fills the universe, we don't pretend to understand it based on "other ways of knowing" that don't involve science. There is tantalizing evidence for dark matter, but we won't claim to know what it is until we have hard evidence. That is precisely the opposite of how the faithful approach their own claims of truth.

In the end, religious investigations of "truth," unlike those of science, are deeply dependent on confirmation bias. You start with what you were taught to believe, or what you want to believe, and then accept only those facts that support your prejudices. This is the basis for the theological practice of "apologetics," designed to defend religion against counterarguments and disconfirming evidence. The fact of evolution, for instance, was once seen by many as strong evidence against God. As we'll see, apologists have now decided that it is exactly what we'd expect from a good creator, who would, of course, allow life to blossom gradually instead of producing a boring and static creation ex nihilo. In contrast, science has no apologetics, for we test our conclusions by *trying* to find counterevidence.

The difference in methodology between science and faith involves several opposing practices and attitudes.

Faith

The most important component of the incompatibility between science and religion is religion's dependence on faith, a word defined in the New Testament as "the substance of things hoped for, the evidence of things not seen." The philosopher Walter Kaufmann characterized it as "intense, usually confident, belief that is not based on evidence sufficient to command assent from every reasonable person." Because Kaufmann was an atheist, we might seek a more neutral definition by going again to the *Oxford English Dictionary*, which gives this "theological" definition of faith:

> Belief in and acceptance of the doctrines of a religion, typically involving belief in a god or gods and in the authenticity of divine revelation. Also (*Theol.*): the capacity to spiritually apprehend divine truths, or realities beyond the limits of perception or of logical proof, viewed either as a faculty of the human soul, or as the result of divine illumination.

Note that what promotes the acceptance of religious doctrine are revelation, "divine illumination," and spiritual apprehension, leading to acceptance of "realities beyond the limits of perception or of logical proof." This is in fact quite similar to Kaufmann's definition, for surely the apprehension of truths that lie beyond normal perception and logic are *not* sufficient to convince most people.

Theologians intensely dislike the definition of faith as belief without—or in the face of—evidence, for that practice sounds irrational. But it surely is, as is any system that requires supporting a priori beliefs without good evidence. In religion, but not science, that kind of faith is seen as a virtue.

If you doubt the claim of my Lutheran debate opponent that faith is a virtue (and the concordant implication that reason is overrated), you can find ample evidence in the works of Christianity, both scriptural and exegetical. Doubting Thomas, who insisted on thrusting his hands into Christ's wounds, was seen as misguided: as Jesus remarked, "blessed are they that have not seen, and yet have believed." Paul and the early church fathers and theologians were unrelenting in their attacks on reason, a doctrine encapsulated in *fideism*, the view that faith and reason are not only incompatible, but also mutually hostile, and that religious belief must be justified by faith alone. Fideism embodies the incompatibility—nay, the war—between science and religion, and is embodied in these two passages, the first from the New Testament and the second from Tertullian (Kierkegaard had similar sentiments):

> But the natural man receiveth not the things of the Spirit of God: for they are foolishness unto him: neither can he know them, because they are spiritually discerned.

> The Son of God died: it is immediately credible—because it is silly. He was buried, and rose again: it is certain—because it is impossible.

It might seem bizarre to believe in something *because* it is absurd, but it makes a kind of sense: faith is required for belief only when you lack good reasons for that belief. Fideism sometimes reaches Orwellian proportions, as it did with Saint Ignatius Loyola:

> To be right in everything, we ought always to hold that the white which I see, is black, if the Hierarchical Church so decides it, believing that between Christ our Lord, the Bridegroom, and the Church, His Bride, there is the same Spirit which governs and directs us for the salvation of our souls.

The view of freethought and curiosity as inferior to faith and religious authority continues today. Although Pope Francis is celebrated for bringing a new spirit of tolerance and modernity to the Vatican, in November 2013 he denigrated the "spirit of curiosity" in a homily at Mass:

> The spirit of curiosity is not a good spirit. It is the spirit of dispersion, of distancing oneself from God, the spirit of talking too much. . . . [And Je-

sus also] tells us something interesting: this spirit of curiosity, which is worldly, leads us to confusion. . . . The Kingdom of God is among us. . . . [Do not] seek strange things, [do not] seek novelties with this worldly curiosity. Let us allow the Spirit to lead us forward in that wisdom, which is like a soft breeze. This is the Spirit of the Kingdom of God, of which Jesus speaks. So be it.

This is a strange attitude given that the Vatican has an astronomical observatory run by priests, complete with a large telescope.

The abnegation of reason is not unique to Catholicism. Martin Luther, for instance, was famous for his many vehement claims that reason was incompatible with Christianity itself. Here are but two:

> For reason is the greatest enemy that faith has: it never comes to the aid of spiritual things, but—more frequently than not—struggles against the Divine Word, treating with contempt all that emanates from God.

> There is on earth among all dangers no more dangerous thing than a richly endowed and adroit reason. . . . Reason must be deluded, blinded, and destroyed.

Believers have a response to the accusation of discerning truth by faith alone. It's the *tu quoque* gambit, which goes something like this: "Well, scientists have faith too: faith in the results produced by other scientists, faith in the empiricism and reason that yield those results, and faith in the idea that it's good to find out more about the universe." We can restate this more simply as: "In these ways, science is just as bad as religion." As we'll see in chapter 4, this claim is false because the meaning of "faith" differs between religious and conventional use.

Authority as the Arbiter of Truth

The dependence on authority is an important difference between science and faith. In many religions, either church dogma or theologians are the final arbiters of truth, and while the flock may deviate from church doctrine, they are not free to concoct their own. "Blasphemy" and "heresy" are terms of religion, not science. A Catholic who rejects the Trinity, for instance, has no power to sway the Vatican's interpretation, and may in fact be excommunicated. The Lutheran theologian whom I debated in Charleston abides by a "confession of faith," which includes three creeds (including the Nicene Creed) as well as the Book of Concord, a compilation of writings by Luther and others. Newly ordained ministers of the liberal Evangelical Lutheran Church in America, for instance, must swear to uphold and promulgate the tenets of that confession, which include the reality of original sin, the virginity of Mary, the Resurrection of Christ, the prerequisite of baptism for gaining eternal life, and the truths of salvation in heaven and eternal punishment in hell.

Now imagine if science worked that way. Upon getting my Ph.D. in evolutionary biology, I'd have to lay my hand on the *Origin of Species* and swear fealty to Darwin and his ideas. The idea is laughable, for such unbending adherence would quickly put an end to scientific progress. Neither scientific texts nor scientists themselves are considered inerrant. Indeed, although I view the *Origin* to be the greatest science book of all time, it's wrong in many respects, including its errors about genetics and about my own area of research—ironically, the origin of species. If scientists were to swear to anything, it would be to abjure all authority in our search for truth.

True, scientists do have confidence (not faith) in some authorities, but only those authorities who have earned trust through a record of either making correct predictions or producing verified observations or experiments. This ethos is embodied in the Latin motto of London's Royal Society, the United Kingdom's most elite body of scientists, physicians, and engineers: *Nullius in verba.* Roughly translated, that means, "Don't take anyone's word." The society notes that this is "an expression of the determination of Fellows to withstand the domination of authority and to verify all statements by an appeal to facts determined by experiment."

It's not widely appreciated that much religious dogma, especially in Christianity, wasn't even derived from scripture or revelation, but from a consensus of opinion designed to quell dissent within the church. The Council of Nicaea, for instance, was convened by Emperor Constantine in 325 to settle issues about the divinity of Jesus and the reality of the Trinity. Despite some dissent, both issues were affirmed. In other words, issues of religious truth were settled *by vote*. The requirement for absolving sin through individual confession wasn't adopted by Catholics until the ninth century; the doctrine of papal infallibility was adopted by the First Vatican Council as late as 1870; and the bodily assumption of Mary into heaven, something debated for centuries, didn't become Catholic dogma until Pope Pius XII declared it so in 1950. And it was only in 2007 that Pope Benedict XVI, acting on the advice of a commission convened by his predecessor, declared that the souls of unbaptized babies could now go to heaven instead of lingering in limbo. Given the absence of new information that produced these changes, how can anybody seriously see this as a rational way to decide religious "truth"?

In fact, changes like the elimination of limbo don't come from new information, but from secular currents in society that make church dogma seem insupportable or even barbaric. The idea of hell, for instance, has become morally repugnant, and so has morphed from an underground barbecue into a more temperate "separation from God."

Just as many churches don't want to be seen as rejecting science, neither do they wish to lag too far behind public morality, and so they often tweak their religious "truths" to reflect the zeitgeist. Perhaps the most obvious example is the Mormon policy on blacks in the priesthood. Although African Americans were permitted to be priests until 1852, church president Brigham Young barred them from the priesthood (Mormons have a lay priesthood) and from participating in certain religious ceremonies—all because their pigmentation supposedly reflected the mark of Cain. A century later this doctrine came under pressure from the American civil rights movement, as well as from the church's desire to make converts in Brazil, a nation where its racial policy would be particularly odious. This produced a timely revision in 1978, ten years after the death of Martin Luther King Jr., when the governing body of the church declared that God, having heard their prayers, sent them a "revelation" giving blacks full privileges as Mormons.

Does anybody believe this account besides Mormons themselves? It's simply too convenient, too opportunistic. And why didn't God know from the outset that discrimination against black Mormons was wrong? What we see here, which holds for many religious doctrines, is "truth" arising not from observation or even revelation but from collusion.

Despite the claims of creationists, who see evolutionary theory as a similar collusion among scientists, there are no such conspiracies in science. The chemist Peter Atkins correctly observed, "Natural selection was a revolution and a stepping-stone to fame; so was relativity, and so was quantum theory. The sheer thrill of discovery is the spur to greater effort. All young scientists aspire to revolution." The same can't be said for theologians (Martin Luther is a rare exception), who either bear their heresies in silence or aspire only to trivial reinterpretations of church doctrine.

Falsifiability

I've already noted that a good scientific theory should be one capable of being shown wrong. When sufficient evidence accumulates that makes a theory no longer tenable, that theory is either altered or discarded.

The attitude toward evidence differs in religion. It's important to note at the outset that religion doesn't automatically *reject* scientific evidence. If we could find nonbiblical confirmation, for example, for the Crucifixion of Jesus, or non-Quranic confirmation for the existence of djinn—the disembodied evil spirits that interact with Muslims—religion would be the first to tout that evidence. If prayers to Allah (but not the Christian God) reliably worked, if amputees or the eyeless could be healed by visiting Lourdes, if archaeological evidence was found for the Exodus from Egypt, religion would proclaim the scientific evidence from the mountaintops, just as would the Christians who still seek the remnants of the Ark on Turkey's Mount Ararat. For decades creationists have tried to show how science supports the story of Genesis.

Where religion diverges from science is how its adherents behave when the evidence *doesn't* support their beliefs. In some cases they behave rationally, and abandon those beliefs—though we shouldn't forget that 64 percent of Americans and 41 percent of British Christians would ditch the science instead. But certain core doctrines are off-limits, and have been immunized against disproof by the construction of a watertight theological edifice—apologetics.

Take the Resurrection of Jesus, for which the only supporting evidence is the contradictory accounts of the Gospels. But suppose we could get evidence *against* it—say, the discovery of ancient texts that tell of a Jesus who didn't revive? It wouldn't matter. Several prominent believers have proclaimed with finality that nothing—nothing—could shake their belief in this and other fundamental claims of Christianity. Here's the prominent theologian William Lane Craig:

> And therefore, if in some historically contingent circumstances the evidence that I have available to me should turn against Christianity, I don't think that that controverts the witness of the Holy Spirit. In such a situation I should regard that as simply a result of the contingent circumstances that I'm in, and that if I should pursue this with due diligence and with time, I would discover in fact that the evidence—if I could get the correct picture—would support exactly what the witness of the Holy Spirit tells me.

Justin Thacker, a theologian at Cliff College, agrees:

> Let's take the resurrection of Jesus Christ. If science somehow, and I can't even imagine how, but if it told me that the resurrection of Jesus Christ was just categorically impossible, could not happen, I would disbelieve that and continue to believe what the Bible teaches about the resurrection of Jesus Christ, because if you take away the resurrection there is no Christian faith, it just doesn't exist.

These statements are, to speak plainly, irrational. Thacker, for instance, deems the Resurrection immune to disproof not because it's supported by strong evidence, but because its absence would undermine his religion. Craig is convinced that with sufficient mental contortions, he'd manage to save his beliefs despite their refutation.

Finally, the liberal Catholic theologian John Haught, who has argued strongly for a harmony between science and religion, has also claimed that if one could have put a camera in Jesus's tomb, it would not have captured an image of Jesus's revival, adding that "if you ask me whether a scientific experiment could verify the Resurrection, I would say that such an event is entirely too important to be subjected to a method which is devoid of all religious meaning."

Now, this is simply a devious way to make an empirical claim but shield it from the possibility of being refuted. Imagine if a cosmologist pronounced the Big Bang too important to be tested!

I haven't cherry-picked these responses while ignoring dissenting views: I've simply never seen any Christian avow in print that he'd abandon belief in the Resurrection if science proved it wrong. Of course, such evidence would be difficult to get, but because the only evidence for the Resurrection is the Bible, which is known to be unreliable in many other matters, it seems judicious to avoid defending Jesus's revival so strongly.

Earlier we heard from Karl Giberson, the evangelical Christian physicist who argued, "Religion often makes claims about 'the way things are.'" Nevertheless, Giberson sees those claims as unfalsifiable, constituting a Procrustean bed into which any annoying facts must be fitted:

> As a believer in God, I am convinced in advance that the world is not an ac-
> cident and that, in some mysterious way, our existence is an "expected" re-
> sult. No data would dispel it. Thus, I do not look at natural history as a source
> of data to determine whether or not the world has purpose. Rather, my ap-
> proach is to anticipate that the facts of natural history will be compatible
> with the purpose and meaning I have encountered elsewhere. And my un-
> derstanding of science does nothing to dissuade me from this conviction.

There can be no clearer statement about the different ways science and religion approach matters of fact. By asserting at the outset that your mind is closed to facts bearing on your beliefs—by asserting that you're *using* faith to buttress your beliefs—you're admitting that you're behaving irrationally. One hopes that Giberson doesn't practice his physics the way he practices his religion.

Cherry-Picking Your "Truths" from Scripture or Authority

Perhaps the most common criticism of prominent atheists like Richard Dawkins is that they see *all* religion as being literalist, and their criticisms are attacks on straw men, comprising at best only a small fraction of believers. This is invariably accompanied by the assertion that literalism is unwarranted because "the Bible isn't a book of science." The arguments come from scientists like Francisco Ayala:

> Genesis is a book of religious revelations and of religious teachings, not a treatise of astronomy or biology.

From liberal theologians like Langdon Gilkey:

> As we have seen, religious explanations are based on special sorts of experience, special insights or revelations, not objective, sharable experiences. Religious theories or beliefs cannot, therefore, be falsified by evidence or by new evidence.

And even from the pope himself:

> How should we understand the narratives of Genesis? The Bible is not intended as a manual of the natural sciences; it wants to help us understand the authentic and profound truth of things.

As I've noted, these are actually claims that the Bible doesn't give us facts. Unfortunately, many believers think it does, including the 30 percent of Americans who regard the book as the actual word of God, and at least some of the 49 percent who see it as a work of humans *inspired* by God. And, of course, most Muslims see the Quran as literally true—and some as a science textbook as well. We'll see later that accommodationist Muslims often argue that the Quran makes truth claims that are perfectly consonant with the findings of modern science.

Sometimes it seems that scriptural literalists are more intellectually honest than the "scripture is not a textbook" crowd, who, rather than admit that science has falsified much of the Bible—and, by implication, has cast doubt on the rest of it—argue that the book is effectively one long parable. After a stiff dose of pick-and-choose apologetics, the words of the Australian creationist Carl Wieland seem like a gust of fresh air:

The Bible's prime purpose certainly concerns salvation, not scientific explanation. But to use this to evade the clear teaching of origins in the foundational book of Genesis is intellectually illegitimate, if not dishonest. . . . Even though the Bible's purpose is not to teach history as such, the history it teaches is true. It states that Jesus was crucified at a specific moment in real history *via* a specific person, Pontius Pilate, the Roman governor of Judea. It would be bizarre to claim that it didn't matter whether these events were true or not, "because the Bible's not a textbook."

The account of Jesus rising from the dead cannot be classified as only one form of truth; i.e. it cannot be a Christian or "religious" truth without at the same time being a historical truth (unless language loses all of its meaning); and it cannot be *historically* true unless it is also *scientifically* true.

The problems with cherry-picking from scripture are obvious. First, we can't step into the minds of the Bible's authors to see what they meant when they wrote it. But we can say this: except for items like the parables of Jesus, clearly meant to be fictitious tales with a moral point, there's no hint in the Bible (or the Quran) that its stories and claims about issues like heaven and hell were meant as anything other than literal truths. That's not to say that these factual claims weren't also meant to convey moral or spiritual lessons, but that those lessons must be discerned from things *that really happened*. In that sense, as Wieland argues, the Bible is indeed a science book, if by that you mean "a book that makes claims about our universe that are really true." And that is how the Bible was interpreted by many, both laypeople and theologians, for millennia—and how the Quran is still interpreted. If the Bible was intended as *pure* allegory, that fact somehow escaped the notice of churches and theologians for centuries.

And if you wed yourself to metaphor rather than fact, how do you know which interpretation is correct? With a bit of imagination—after all, there's no guide to truth here—you can pick almost any biblical story, say, that of Adam and Eve, and discern several competing explanations. In fact, even as I write, theologians, annoyed by the genetic disproof of Adam and Eve as our literal ancestors, are engaged in mental gymnastics trying to find metaphors in a story that has been falsified. The story of Job has baffled scholars for centuries, for its "meaning" is murky, yet there is no lack of those willing to give it a metaphorical spin.

The big problem for believers, of course, is to find a consistent method for distinguishing fact from metaphor. If Adam and Eve are metaphors, could the Resurrection be a metaphor as well—perhaps for spiritual rebirth? And how are we supposed to interpret God's commands in the Old Testament that a man gathering sticks on the Sabbath should be put to death, as should practicing homosexuals, adulterers, and those who curse their parents? Because these are no longer seen as valid moral prescriptions, are they really metaphors for something else, or did God simply change his mind?

The way liberal believers winnow fact from metaphor is in fact to use science: whatever science has falsified becomes metaphor, and what has not yet been falsified can retain its status as fact. But this makes religious dogma subservient to science. The accommodationist strategy of accepting both science and conventional faith, then, leaves you with a double standard: rational on the origin of blood clotting, irrational on the Resurrection; rational on dinosaurs, irrational on virgin births. Scientists, archaeologists, and historians can tell us which parts of scripture are false, but who can affirm what is true? There are no good criteria.

Turning Scientific Necessities into Theological Virtues

This tactic goes beyond converting a false religious claim into a metaphor: the metaphor is further turned into a *virtue*. In other words, by correcting scripture, science gives us a theology that's even better than we were able to glean from the Bible in prescientific days.

Evolution is the prime example of how the theological sausage grinder can transform scientific necessities—empirical findings that contradict scripture but convince more rational believers—into religious virtues. The apparent "design" of plants and animals was once the centerpiece of "natural theology," the discipline that tries to find evidence for God and his characteristics from studying nature. Before 1859, there simply was no alternative to God's ingenuity as the explanation for the remarkable adaptations of animals and plants: the spiny, water-conserving shapes of desert plants, the cryptic coloration of

flounders and chameleons, and the aerodynamic skin flaps of the flying squirrel. But all that changed when Darwin explained those designlike features by
natural selection. The best evidence for God simply vanished.

What did apologists do when the Genesis story was so thoroughly
trounced? They made its refutation a virtue, arguing that it was notably *better* for God to have created through evolution than by poofing life into existence like a magician. Evolution is, after all, supposedly contingent and
unpredictable, allowing God a form of creativity unavailable if organisms
were created from nothing.

Both scientists and theologians have offered this rationalization, including the evolutionary biologist Francisco Ayala, formerly a Dominican priest:

> A world of life with evolution is more much exciting; it is a creative world
> where new species arise, complex ecosystems come about, and humans
> have evolved.

The geneticist and physician Francis Collins admires God's ingenuity at
using evolution to produce our own species:

> Seeking to populate this otherwise sterile universe with living creatures,
> God chose the elegant mechanism of evolution to create microbes, plants,
> and animals of all sorts. Most remarkably, God intentionally chose the
> same mechanism to give rise to special creatures who would have intelli-
> gence, a knowledge of right and wrong, free will, and a desire to seek fel-
> lowship with Him. He also knew these creatures would ultimately choose
> to disobey the Moral Law.

It's obvious that if you can justify every scientific advance as comporting
perfectly with God's will, then science can never refute the claims of your
faith. And claims that can't be refuted can't be confirmed.

Fabricating Answers to Hard or Insoluble Questions

One can't read a great deal of theology without appreciating the mental dexterity of its practitioners when faced with hard problems. And as a scientist, I

regret how much more we'd understand about nature had that dexterity been applied to science instead, or to any field involved in studying what's real.

Take, for instance, the question of why God is hidden. Even if you see the Bible as largely allegorical, God's presence—revealed by miracles, resurrections, and the like—was far more evident two millennia ago than today. And despite the "miracles" of Lourdes and Fatima, more sophisticated believers must deal with the issue of the *Deus absconditus*, the hidden God. The most parsimonious hypothesis is to simply claim that there are no gods, so their absence is expected. But that's unacceptable to the religious, who are then obliged to respond. Here's one answer from John Polkinghorne and his collaborator, the social philosopher Nicholas Beale:

> The presence of God is veiled because, when you think about it, the naked presence of divinity would overwhelm finite creatures, depriving them of truly being themselves and freely accepting God.

But does that really make sense? Would *you* find the naked presence of God so overwhelming that you'd reject him? That certainly wasn't the case when he appeared to Moses as a burning bush or to Job as a whirlwind. One would think that "finite creatures" would be delighted with tangible evidence for their beliefs.

The theologian John Haught has a different answer:

> It is essential to religious experience, after all, that ultimate reality be beyond our grasp. If we could grasp it, it would not be ultimate.

This wordplay not only misuses the term "ultimate," but is deeply tautological, giving no answer at all.

Some liberal believers will reluctantly admit that God's absence makes them uncomfortable. Surprisingly, one of them is Justin Welby, the current archbishop of Canterbury, who acknowledged his doubts about whether God existed at all. In an interview with the BBC at Bristol Cathedral, he confessed, "The other day I was praying over something as I was running and I ended up saying to God: 'Look, this is all very well but isn't it about time you did something—if you're there'—which is probably not what the archbishop of Canterbury should say." But he hastened to add that those doubts didn't extend to Jesus, whose existence, Welby claimed, was a dead certainty: "We know about Jesus, we can't explain all the questions in the world, we can't explain about suffering, we can't explain loads of things but we know about Jesus." How one can be certain about the divinity of Jesus but not about God's existence escapes me. It's like saying, "I'm not so sure about Santa Claus, but I've no doubts about his reindeer." (Indeed, Anglicans accept the Holy Trinity—the "Unity of the Godhead"—so Jesus and God are part of the same entity.) And so the liberal religious mind treats claims differently, even if there is equally little evidence supporting them.

In science, if there *should* be evidence for a phenomenon but that evidence is consistently missing, one is justified in concluding that the phenomenon doesn't exist. Examples are the Loch Ness Monster and Bigfoot, as well as paranormal phenomena like ESP and telekinesis. Seeking evidence for such things, the skeptics always come up dry. It is the same with God, though theologians will object to comparing God to Bigfoot. The philosopher Delos McKown had a more parsimonious answer for God's absence: "The invisible and the non-existent look very much alike."

What about immortality in an afterlife? Many ardently wish for it, but the evidence is annoyingly scarce. Given that nobody has returned from the dead, and that anecdotes involving reincarnation or near-death visits to heaven are either dubious or conflicted, or have mundane physiological explanations, how do we reassure ourselves of eternal life? One way is to simply reject the *need* for evidence, as does John Haught:

> In any case, were I to try to elicit scientific evidence of immortality I would just be capitulating to the narrower empiricism that underlies naturalistic belief. What I will say, though, is that the hope for some form of subjective survival is a favorable disposition for nurturing trust in the desire to know. . . . Such a hope is reasonable if it provides, as I believe it can, a climate that encourages the desire to know to remain restless until it encounters the fullness of being, truth, goodness and beauty.

Note the cavalier dismissal of "narrow empiricism"—that is, real evidence. But either immortality exists or it doesn't, and surely such an important

matter must rest on more than "hope for some form of subjective survival."
How else can one satisfy the "desire to know" without knowledge itself?

If apologetics is to be a satisfying system of explanation, it has to deal
with the "hard problem" of theology: natural evil. While *moral* evils, like
thefts and murders, are often justified as the unavoidable by-product of
God's gift of free will, there is no obvious explanation for those tragedies,
like childhood cancers and deadly earthquakes, that inflict suffering on the
innocent and undeserving. This is not the place to rehash the diverse and
creative theological explanations for how natural evil comports with a loving and all-powerful God, though I find none of them remotely convincing
(was the Holocaust really necessary to preserve the free will of Germans?).
But science does better, for such evils are precisely what one expects in a
purely naturalistic world. Tumors in children? The result of random mutations. Tsunamis and earthquakes? The endless churning of the Earth's crust
produced by plate tectonics. The horrible suffering inherent in evolution by
natural selection? The inevitable result of competition among genes that
control the bodies of living creatures.

It was in fact the suffering produced by natural selection—and the death
of his beloved ten-year-old daughter Annie—that helped wean Darwin
from religion. In a letter to the American botanist Asa Gray written only six
months after he published *On the Origin of Species,* Darwin pondered the
disparity between traditional theism and the suffering of animals:

> With respect to the theological view of the question; this is always pain-
> ful to me.—I am bewildered.—I had no intention to write atheistically.
> But I own that I cannot see, as plainly as others do, & as I shd wish to do,
> evidence of design & beneficence on all sides of us. There seems to me too
> much misery in the world. I cannot persuade myself that a beneficent &
> omnipotent God would have designedly created the Ichneumonidæ with
> the express intention of their feeding within the living bodies of caterpil-
> lars, or that a cat should play with mice.

Although Darwin was usually silent on the conflict between his theories
and religion (his wife, Emma, was quite devout), I see in this letter a feeling
that his own theory supersedes any religious explanation for suffering.

*Applying Different Standards to One's Own Religious Beliefs
Than to Those of Other Faiths*

Here are some truth claims of different religions, taken from their theology:

Seventy-five million years ago, Xenu, the dictator of a galactic confederation, brought billions of humanlike beings to Earth in a giant spaceship that resembled a Douglas DC-8. Paralyzed and then preserved in antifreeze, their bodies were piled up around the bases of volcanoes and destroyed by exploding hydrogen bombs within the craters. Their escaped souls, called "thetans," were captured, taken to a giant cinema, and forced to watch movies for about a month, implanting in the thetans bad ideas like Catholicism. The thetans then escaped, affixing themselves to the bodies of those who survived the explosions. Humans afflicted with thetans can be diagnosed only with special devices that measure skin conductance.

You don't believe that, right? But that is official doctrine of the Church of Scientology, concocted by the science-fiction writer L. Ron Hubbard and seen as gospel by his church and its adherents, many of them paying thousands of dollars to learn these "truths." If you don't accept that story, why not?

How about this one?

In 1827, New Yorker Joseph Smith, guided by an angel named Moroni, unearthed a binder of golden plates written in strange characters. With the help of his hat and a "seer stone," Smith translated the plates into English. This transcription, the Book of Mormon, claims that Jesus visited North America and that Native Americans are the descendants of people from the Middle East who migrated to North America.

If that doesn't seem credible—and it does to the faith's fifteen million adherents—remember that the Book of Mormon begins with the sworn testimony of eleven witnesses who claimed to have seen the plates. These were actual, living people—giving the Book of Mormon far more historical credibility than the Bible or the Quran.

Here's another:

Illness and deaths are illusions—purely the result of faulty thinking—and even "diseases" like diabetes and cancer can be overcome by proper belief.

That's the doctrine of the Christian Science church, founded in 1879. As we'll see, hundreds of people have died relying on this theology rather than receiving proper medical care.

Most of the world's believers reject these claims as blatantly false. But that's only because these three religions are fairly new. They were founded in the last two centuries, and we see their origin not as divine but as obvious fabrications of humans—in the case of Joseph Smith, of a con man. But if you look with equally critical eyes at the doctrines of older faiths, their tenets seem equally bizarre. Islam, for instance, claims that Muhammad was accosted by two angels who split open his breast, extracted his heart, and purified it with snow, rendering him suitable to be God's prophet. The angel Gabriel then commanded him to recite, which he did for twenty-three years, producing the Quran. And, of course, the Christian mythology includes stories of talking serpents, worldwide floods, virgin births, and a divine prophet who, after resurrecting the dead and healing the blind, was resurrected himself. The obvious question is this: why are believers in mainstream religions, like Islam and Christianity, less critical of their own faiths than of others?

One reason is that most mainstream faiths have been around for millennia. Because we weren't there when they were founded, we can't dismiss their divine origins as readily as we can for Scientology or Mormonism. Their persistence has given them an aura of credibility, somehow making their claims seem less contrived.

But the main reason people turn a blind eye toward implausible beliefs is that they get their faith not through reason or deliberation, but through indoctrination from their family and friends. Religion has hijacked the evolved tendency of humans to accept authority when they're young, something that would have enhanced the survival of our ancestors (learning is a good way to avoid the dangers of experience). And so if you're born in Saudi Arabia, in all likelihood you'll be brought up Muslim, accepting its doctrines as true. If born in Utah, the chances of your becoming a Mormon are high (around 60 percent), and in Brazil you're likely to become a Catholic. To a very large extent, which religion you accept and which you reject are accidents of birth. And after you've been religious for years, and surrounded by those who believe likewise, you become emotionally invested in your faith's truth. This makes you more susceptible to confirmation bias and less likely to be skeptical about your beliefs.

But none of these are good reasons for deeming your religion true and others false. Granted, some people argue that *all* religions are true, claiming that at bottom we all worship the same God. But that's simply not the case. The foundational claims of different religions are not only disparate, but conflicting. Many Christians think that the only route to salvation is accepting Jesus as one's savior. If you're a Muslim, that doctrine will send you straight to hell. The Quran also claims that Jesus was slain but not crucified, with an impostor dying on the cross. Jews, of course, don't see Jesus as the Messiah at all.

Unlike monotheistic faiths, Hinduism has many gods. Jehovah's Witnesses think that precisely 144,000 of them will make it to heaven, while the others who are saved will inhabit a paradise on Earth. In contrast, Laestadianism, a conservative branch of Lutheranism, considers itself the only true faith: only its roughly sixty thousand adherents are eligible for salvation, with the billions of others on Earth doomed to eternal torment. Catholics believe in *transubstantiation:* that the wine and wafer consumed during the Eucharist actually become the physical substance of Jesus's body and blood. In contrast, some Eastern Orthodox and Protestant sects hold to *consubstantiation,* the notion that the wine and wafer coexist as regular food and drink along with Jesus's blood and body. How could one possibly distinguish between these claims? We'll never know who gets saved, and chemical or DNA tests will show that bread and wine remain bread and wine during *all* Eucharists. What basis, then, for these beliefs? (Remember that the Eucharist involves not a metaphorical and spiritual transformation, but an actual *physical* transformation.)

Even more bizarrely, Black Muslims believe that whites are a race of devils, created less than seven thousand years ago from selective breeding by a mad black scientist named Yakub. And, of course, there is Xenu and his hydrogen bombs. Add to these all the conflicting doctrines and equally conflicting moral codes that differ in how one should treat women, gays, sex before or outside of marriage, criminals, animals, and so on. They can't *all* be right.

How many different religions are there? The number is uncountable. While there are about a dozen "major" religions, they're fractured into different branches with different beliefs and practices. In fact, the Gordon Conwell Theological Seminary estimates that there are forty-four thousand sects of Christianity alone!

The different claims among these faiths have consequences, for they've produced endless misery over the course of history. Sunni and Shiite Muslims, who regularly kill each other, originally diverged only in whom they saw as the proper person to head the faith: those related to Muhammad or those who, regardless of ancestry, were most qualified. "Heretical" Christian sects like the Donatists and Cathars were ruthlessly extirpated over differences in doctrine. And even today, 16 percent of the world's 198 countries penalize blasphemy, while 20 prohibit apostasy (abandoning one's faith). All of the latter are nations that are largely Muslim.

Clearly, religions aren't incompatible only with science: they're incompatible with one another. And this incompatibility wasn't inevitable: if the particulars of belief and dogma were somehow bestowed on humans by a god, there's no obvious reason why there should be more than one brand of faith. These schisms and conflicts are further evidence that religion is not only a human construct, but is about more than sociality or community. *Beliefs matter.*

But suppose that there is a "correct" religion—one whose conception of God, and the practices and moral codes God decrees, is accurate. How do you discover it? Given that most religious people acquire their faith through accidents of birth, and those faiths are conflicting, it's very likely that the tenets of a randomly specified religion are wrong. How can you tell if *yours* is right? As we've seen, this question should be of the greatest importance to believers, for its answer involves the all-important issues of morality and, if you believe in an afterlife, where you'll spend eternity.

The only rational solution is to apply the same degree of skepticism toward the claims of your own faith as you do toward the ones you reject. This rational and quasi-scientific approach is promoted by the ex-preacher John Loftus, who lays it out briefly:

> It is highly likely that any given religious faith is false and quite possible that they could all be false. At best there can only be one religious faith that is true. At worst, they could all be false. . . .
> So I propose that: . . . The only way to rationally test one's culturally

adopted religious faith is from the perspective of an outsider with the same level of reasonable skepticism believers already use to examine the other religious faiths they reject. This expresses the *Outsider Test for Faith* (OTF).

Given that beliefs matter, the wisdom of this approach is unquestionable. But if it's used honestly, its outcome is inevitable. If you're a Christian, for instance, you probably reject the beliefs of Islam because you see them not only as misguided, but also as lacking in evidence. If that's the case, then you must abandon your own faith on the same grounds. In the end, the inconsistencies between faiths, combined with the reasonable doubt that believers apply to other faiths, means that *no* faiths are privileged, none should be trusted, and all should be discarded. This is what the philosopher Philip Kitcher calls the core challenge of secularism toward religion: the "argument from symmetry."

This farrago of conflicting and irresolvable claims about reality stands in stark contrast to science. While science itself has fragmented into different disciplines that use different tools, they all share a core methodology based on doubt, replication, reason, and observation. In other words, while there are different sciences, there is only one *form* of science, whose conclusions don't depend on the ethnicity or faith of the scientist who reaches them. Because of this, we need no Outsider Test for Science.

Scientific Truth Is Progressive and Cumulative; Theological "Truth" Isn't

The progress of science is palpable to everyone, whether you measure it in improvements in the quality or length of our lives (the average life span has doubled since 1800), or simply in our improved understanding of nature. During my own lifetime I've seen the elimination of smallpox and the virtual elimination of polio, the discovery of the Big Bang, the uncovering of the structure of DNA and how it produces bodies, the ability to transplant organs, the reconstruction of much of the evolutionary history of life, the advent of personal computers, the first Moon landing and space shuttle, the sending of Mars rovers to explore the planet, in vitro fertilization, cell phones, HPV vaccines, and the identification of the Higgs boson. And that's just since 1949.

Has theological knowledge advanced since 1949? Clearly not. In fact, I would argue that it hasn't advanced in the last five thousand years, more than ten times longer than the span of modern science.

Now, when I say theology hasn't advanced, I'm not saying that it hasn't *changed,* for it clearly has. The idea of hell has been abandoned by many, or reconceived as a "separation from God." The notion of the Immaculate Conception has been adopted. The Catholic Church eliminated (in 1966!) the *Index Librorum Prohibitorum,* the list of banned books considered injurious to morality, including works by Hume, Sartre, Milton, Locke, and Copernicus. Surely the ability to read these authors without damnation can be seen as an advance—but not an advance in theology. And, of course, whole branches of theology have sprouted, including process theology and liberation theology, as well as wholly new religions, like Scientology and Christian Science.

Nor am I saying that some aspects of church doctrine haven't come into better sync with reality. The old views of Adam and Eve, of an instantaneous creation, and of a worldwide flood have largely fallen out of favor because science has shown them to be false. Finally, I am not claiming that the *morality* advanced by some religions hasn't advanced, for it has: many churches now espouse gay rights and women's rights, and, on the whole, Western religion has become more enlightened and liberal. Liberation theology is in fact a movement designed to infuse traditional Catholicism with notions of social justice. You'd be hard pressed to find a church that still supports slavery, though many did before the Civil War.

What I am saying is two things. First, religion hasn't obviously come closer to understanding the divine. From the ancient Hebrew sages through Aquinas to Kierkegaard, we still have no idea whether gods exist; whether there is only one god or many; whether any existing god is deistic, and largely absent, or theistic, interacting with the world; what the nature of any god is (is it apathetic, kindly, or evil?); whether that God is, as process theology claims, affected and changed by the world or unchangeable; how God wants us to live; and whether there is an afterlife, and, if so, what it is like. What has happened is that new theologies and religions have simply appeared alongside the old ones, with some of the old ones going extinct. We still have Judaism, but now we also have Catholicism, Mormonism, and Scientology. Ancient polytheistic Hinduism is still here, side by side with Buddhism and aboriginal religion. Fundamentalists, who see almost the entire Bible as literal truth, coexist with apophatic theologians who claim that nothing can be said about God, and yet write many books on the topic. In this way theology is not progressive but additive, and no consensus has developed about gods and their will. This, of course, contrasts strongly with science, where consensus views have evolved in every field—views that may change with time, but always lead to a deeper understanding of the universe, one that expands our abilities and makes our predictions more accurate. Before 1940, there was no way to decide which primate was our closest relative, to land rockets on the Moon, to understand how the genetic material coded for bodies and behavior, or to determine our position on the planet within a few meters.

I also claim that insofar as theology or religious beliefs *do* change within a faith, those changes are driven largely by either science or changes in secular culture. It is science, of course, that has put paid to most of the creation myths in Genesis, and archaeology to myths like the Exodus and the captivity of the Jews in Egypt. And it is largely advances in secular philosophy, like increasing empathy for minorities and the dispossessed, that have fueled changing notions of hell, the infusion of social justice into churches, and the acceptance of minorities and women. Religious morality, at least as promulgated by priests, rabbis, imams, and theologians, is usually one step behind secular morality. We are seeing this play out at the moment as increasing numbers of Catholics take issue with church dogma about abortion, contraception, male priests, and homosexuality. We can be fairly confident that eventually the church will make some moral concessions. Like biological species, churches must adapt or die when their environment changes radically.

In candid moments, some theologians like John Polkinghorne grudgingly admit that theology moves more slowly than science:

The nesting relationship of successive scientific theories gives the subject its character of a cumulative advance of knowledge. A very ordinary sci-

entist today possesses, in consequence, much greater overall understanding of the physical world than was ever possible for Sir Isaac [Newton]. . . .
The theologian of the twentieth century enjoys no presumptive superiority over the theologians of the fourth or sixteenth centuries. Indeed, those earlier centuries may well have had access to spiritual experiences and insights which have been attenuated, or even lost, in our own time.

But even here theology can turn necessity into virtue. The philosopher and theologian J. P. Moreland sees theology's stasis as a sign of its ability to grasp truth *more readily* than does science!

The slow progress in philosophy and theology may indicate not that they are less rational than science—that is, that they have progressed less toward truth—but that they are more rational. Why? Because the slow progress could be an effect of their already having eliminated proportionately more false options in their spheres of study than science has eliminated in its. If this is true, it means that they have already come closer to a full, well-rounded true world view than science has come.

In sum, philosophy and theology may not progress because they may have already arrived rationally at some truth concerning the world. This means that a philosopher or theologian has the right to be sure about this conclusion, not in the sense of terminating inquiry or being closed to new arguments, but in the sense of requiring a good bit of evidence before abandoning the conclusion and not being able to use it to infer other conclusions.

This is a textbook example of how to rationalize uncomfortable truths. But if Moreland is right, let him tell us which of the thousands of religious worldviews is "true" and which have been discarded as false.

The methodological conflicts between science and religion cannot be brokered, for faith has no reliable way to find truth. It is no more compatible for someone to be a scientist in the lab and a believer in church than it is for someone to be a science-based physician who practices homeopathic medicine in her spare time.

Conflicts of Outcome

While most believers accept the methods of science, they also claim additional methods of apprehending truth: faith, revelation, and authority. If those methods were reliable, then the outcomes of religious and scientific investigations would be similar, or at least consonant.

But, of course, they aren't. Religions make truth claims that have been repeatedly disproved, claims involving both natural and supernatural phenomena (I'll say more about the supernatural later). The disproved religious claims involve biology, geology, history, and astronomy, and include these assertions: animals and plants were created in their present form over a short period of time, the Earth is young and was once completely inundated by a great flood, modern humans descend from only two progenitors, Native Americans descend from immigrants from the Middle East, and Caucasians are the results of a breeding experiment by a black scientist. These are all wrong, and it's science, not faith, that has shown them to be wrong.

Even the *historical* claims of religion, not counting the various origin myths, are often dubious. There is, for instance, no evidence for the Exodus of Israelites from Egypt, or for the census of the entire Roman Empire around the time of Jesus's birth as described in the Gospel of Luke. As we've seen, there's no reliable historical account—and there *should* be one—for the Crucifixion miracles, like earthquakes and resurrected saints, described in the Gospel of Matthew. How did historians of the time miss *that*? True, some historical facts adduced in the Bible are accurate, for it was written by people living in those times. There is evidence, for instance, for a Roman governor of Judaea named Pontius Pilate, though no extrascriptural evidence that he judged Jesus. But biblical archaeology has, by and large, experienced one failure after another. If you have no problems rejecting biblical incidents like the Exodus or the census of Caesar Augustus, incidents that, like the Resurrection, come solely from scripture, why accept the Resurrection itself?

For if the checkable truths of religion—the "natural truths"—are faulty, why should we give credence to the harder-to-test "divine truths"? Did those claims—the existence of souls, the birth of Jesus from a virgin and his subsequent execution and Resurrection, the presence of an afterlife, the existence of demons, the ascent of Muhammad to heaven on a wingèd horse—just happen to be the ones that God or his scribes got correct, while they erred on many others? If the Bible can't even get the basic facts of history right, much less those of science, how can we claim divine authority or influence in its authorship? Was it beyond God, speaking either directly or through his emissaries, to tell his creatures that it was advisable to wash their hands after defecating, or that animals and plants weren't created suddenly but evolved from other forms over a long period of time?

Over the years, I've repeatedly challenged people to give me a single verified fact about reality that came from scripture or revelation alone and then was confirmed only later by science or empirical observation. This parallels Christopher Hitchens's moral challenge, often leveled at religious opponents in debates: "Name me an ethical statement or an action performed by a believer that could not have been made or performed by a non-believer." Like Hitchens, I've never had a credible response.

Conflicts of Philosophy

The methodological conflicts between science and religion have ultimately produced a conflict in philosophy: whether or not one sees gods as a realistic possibility. It's important to realize that this philosophical difference between scientists and believers was not established at the outset as an integral part of science, but arose gradually as a by-product of science's success.

Science is now deeply wedded to *naturalism*, the view that all of nature operates according to laws—or rather, "regularities," for the word "laws" implies a lawgiver—and that a combination of theoretical and empirical study can reveal those laws. (A related phenomenon is *materialism*, the view that matter and energy are all there is to the universe. I prefer to use "naturalism" because it's always possible that we'll find some stuff, like "dark matter," that is neither matter nor energy as we understand them now.) One of the main criticisms of science by philosophers and theologians is that scientists are *committed* to naturalism, almost as if we had to swear allegiance to the idea when we got our science degrees. But this criticism is wrong. Naturalism is not something that was always part of science, for at one time science *did* rely on supernatural explanations. As modern science established itself, there was a period when both natural and supernatural explanations abounded, and only gradually did science slip the bonds of the divine. Creationism is one example—the only credible explanation, before 1859, for the remarkable fit of organisms to their environments. But invoking principles other than naturalism never helped us make progress. It was Darwin's reliance on the naturalistic idea of natural (as opposed to supernatural) selection that correctly explained biological adaptation and diversity.

An anecdote about the French mathematician Pierre-Simon Laplace provides the classic example of why we adopted naturalism. The backstory involves the astronomical work of Isaac Newton. Genius that he was, Newton nevertheless invoked God as a scientific hypothesis. He thought, for instance, that the orbits of the planets would be unstable without God's occasional intervention to keep them in place. It was Laplace who later showed that such divine twiddling was unnecessary, and that natural law alone was adequate. The superfluity of religious explanations is described in a story that, though often told, may well be apocryphal. Laplace was said to have given Napoleon Bonaparte a copy of his great five-volume work on the solar system, the *Mécanique Céleste*. Aware that the books contained no mention of God, Napoleon supposedly taunted him, "Monsieur Laplace, they tell me you have written this large book on the system of the universe, and have never even mentioned its Creator." Laplace answered, famously and brusquely, "Je n'avais pas besoin de cette hypothèse-la"—"I have had no need of that hypothesis." And scientists haven't needed it since.

Our reliance on naturalism, then, is not an assumption decided in advance, but a result of experience—the experience of men like Darwin and Laplace who found that the only way forward was to posit natural rather than supernatural explanations. Because of this success, and the recurrent failure of supernaturalism to explain *anything* about the universe, naturalism is now taken for granted as the guiding principle of science. Its use in all scientific studies is called *methodological naturalism* or—because it's used to explain observations—*methodological empiricism*.

Nevertheless, some scientists persist in claiming, wrongly, that naturalism is a set-in-stone rule of science. One of these is my Ph.D. adviser, Richard

Lewontin. In a review of Carl Sagan's wonderful book *The Demon-Haunted World,* Lewontin tried to explain the methods of science:

> It is not that the methods and institutions of science somehow compel us to accept a material explanation of the phenomenal world, but, on the contrary, that we are forced by our *a priori* adherence to material causes to create an apparatus of investigation and a set of concepts that produce material explanations, no matter how counter-intuitive, no matter how mystifying to the uninitiated. Moreover, that materialism is absolute, for we cannot allow a Divine Foot in the door.

That quotation has been promulgated with delight by both creationists and theologians, for it seems to show the narrow-mindedness of scientists who refuse to even admit the *possibility* of the supernatural and immaterial.

But Lewontin was mistaken. We can *in principle* allow a Divine Foot in the door; it's just that we've never seen the Foot. If, for example, supernatural phenomena like healing through prayer, accurate religious prophecies, and recollection of past lives surfaced with regularity and credibility, we might be forced to abandon our adherence to purely natural explanations. And in fact we've sometimes put naturalism aside by taking some of these claims seriously and trying to study them. Examples include ESP and other "paranormal phenomena" that lack any naturalistic explanation.

Sadly, arguments similar to Lewontin's—that naturalism is an unbreakable rule of science—are echoed by scientific organizations that want to avoid alienating religious people. Liberal believers can be useful allies in fighting creationism, but accommodationists fear that those believers will be driven away by any claim that science can tackle the supernatural. Better to keep comity and pretend that science *by definition* can say nothing about the divine. This coddling of religious sentiments was demonstrated by Eugenie Scott, the former director of an otherwise admirable anticreationist organization, the National Center for Science Education:

> First, science is a limited way of knowing, in which practitioners attempt to explain the natural world using natural explanations. By definition, science cannot consider supernatural explanations: if there is an omnipotent deity,

there is no way that a scientist can exclude or include it in a research design. This is especially clear in experimental research: an omnipotent deity
cannot be "controlled" (as one wag commented, "you can't put God in a test
tube, or keep him out of one"). So by definition, if an individual is attempt-
ing to explain some aspect of the natural world using science, he or she
must act as if there were no supernatural forces operating on it. I think this
methodological materialism is well understood by evolutionists.

Note that Scott claims naturalism as part of the *definition* of science.
But that's incorrect, for nothing in science prohibits us from considering
supernatural explanations. Of course, if you define "supernatural" as "that
which cannot be investigated by science," then Scott's claims become tautologically true. Otherwise, it's both glib and misleading to say that God is
off-limits because he can't be "controlled" or "put in a test tube." Every study
of spiritual healing or the efficacy of prayer (which, if done properly, includes controls) puts God into a test tube. It's the same for tests of nondivine supernatural phenomena like ESP, ghosts, and out-of-body experiences.
If something is supposed to exist in a way that has tangible effects on the
universe, it falls within the ambit of science. And supernatural beings and
phenomena can have real-world effects.

In the end, the incompatibility between the methods of science and religion ultimately yields a genuine *philosophical* incompatibility. Working
scientists are constantly steeped in doubt and criticality, leading to an ingrained skepticism about truth claims—not a bad thing, really. If you com-
bine that attitude with the proven value of naturalism over the history of
science, and mix in the repeated failure to find evidence for the supernatural, then you arrive at the following philosophy: because there is no evidence for supernatural entities or powers, although there *could* have been
such evidence, one is justified in thinking that *those entities and powers do
not exist.* This attitude is called *philosophical naturalism.*

The philosopher Barbara Forrest defends the connection between methodological and philosophical naturalism:

Taken together, the (1) proven success of methodological naturalism
combined with (2) the massive body of knowledge gained by it, (3) the

lack of a comparable method or epistemology for knowing the supernatural, and (4) the subsequent lack of any conclusive evidence for the existence of the supernatural, yield philosophical naturalism as the most methodologically and epistemologically defensible world view.

This is where philosophical naturalism wins—it is a *substantive* world view built on the cumulative *results* of methodological naturalism, and there is nothing comparable to the latter in terms of providing epistemic support for a world view. If knowledge is only as good as the method by which it is obtained, and a world view is only as good as its epistemological underpinning, then from both a methodological and an epistemological standpoint, philosophical naturalism is more justifiable than any other world view that one might conjoin with methodological naturalism.

Although Forrest wrongly implies that science can't examine the supernatural, her overall argument makes sense. If you spend your life looking in vain for the Loch Ness Monster, stalking the lake with a camera, sounding it with sonar, and sending submersibles into its depths, and yet still find nothing, what is the more sensible view: to conclude provisionally that the monster simply isn't there, or to throw up your hands and say, "It *might* be there; I'm not sure"? Most people would give the first response—unless they're talking about God.

Some scientists succeed at being methodological naturalists on the job and supernaturalists at other times, but it's hard to reconcile an ingrained skepticism toward the claims of your colleagues with complete credulity toward the claims of fellow believers. The skepticism usually leaks through, explaining both why many scientists become atheists and why many believers instinctively distrust or even disparage science. The hemorrhage of scientific doubt into the body of faith is also, I think, why older scientists are less religious than younger ones (they've been critical for longer), and why more-accomplished scientists are also less religious (their criticality, and willingness to question authority, is what brought them renown).

It's important to realize that philosophical naturalism is, like atheism, a *provisional* view. It's not the kind of worldview that says, "I *know* there is no god," but the kind that says, "Until I see some evidence, I won't accept the existence of gods." Even so, philosophical naturalism is anathema to accommodationists, who, even though they may be atheists themselves, avoid flaunting their beliefs. And even scientists who embrace that philosophy tend to keep quiet about it, for, at least in America, we're surrounded by believers, some of whom fund our research and others who help us fight creationism and pseudoscience. It's not clear whether professing atheism would endanger those alliances, but in America, where an atheist is a skunk in the woodpile, it seems better to play it safe.

The most objectionable thing an American scientist can say about belief is this: "Science and religion are incompatible, and you must choose between them." After all, many people embrace both, and we know that when forced to choose, many would keep their faith. On such grounds accommodationists argue that forcing that choice is both unseemly and damaging to science. But if you are trying to be consistent in how you get reliable knowledge about our universe, and if you already reject unevidenced claims like those of ESP and homeopathy, or claims of religions other than yours, then in the end you *must* choose—and choose science. That doesn't mean that you must accept an unchanging set of scientific facts, but simply that you choose reason and evidence over superstition and wish-thinking.

CHAPTER 3

Why Accommodationism Fails

There is no harmony between religion and science. When science was a child, religion sought to strangle it in the cradle. Now that science has attained its youth, and superstition is in its dotage, the trembling, palsied wreck says to the athlete: "Let us be friends." It reminds me of the bargain the cock wished to make with the horse: "Let us agree not to step on each other's feet."

—Robert Green Ingersoll

"Cognitive dissonance" is a well-known phenomenon in which you experience psychological discomfort from holding two conflicting beliefs or attitudes, or from behaving in a way that is inconsistent with your beliefs. Such a dilemma causes mental discomfort, compelling you to find a way to reduce that distress. A common solution is to convince yourself that there's not really a conflict. People who think of themselves as honest may cheat a bit on their taxes, but then rationalize it, preserving their self-image by saying, "It's not so bad, because *everyone* does it, and anyway, the government wastes a lot of money."

Accommodationism—claiming that science and religion are not in conflict—is the solution to another form of cognitive dissonance, the one that appears when you live in a culture that reveres science but you still cling to pseudoscientific and religious myths. Many people want to be seen as pro-science, but they also need the comfort of their faith. And so they cobble together a variety of solutions that let them have both.

The most visible accommodationists are religious people, especially theologians, who are liberal, friendly to science, but see that scientific inquiry

poses some threat to their beliefs. The dissonance is especially strong in religious *scientists*, whose practice at work directly conflicts with their faith's
"ways of knowing." Surprisingly, though, many of the most prominent accommodationists are atheists or agnostics. Some of these, whom I call "faitheists,"
see religion as a falsehood, but one that's good for society (the philosopher
Daniel Dennett has called this attitude "belief in belief").

There are also political reasons for accommodationism. As I've noted,
American science educators and science organizations coddle faith to gain
religious allies in our fight against creationism, to reassure a religious government (the main supporter of research) that science isn't equivalent to
atheism, and to avoid a reputation as rabble-rousers or, worse, God-haters.
I can't tell you how many times I've heard the claim that publicly professed atheism harms the acceptance of evolution. Here's one example from
Roger Stanyard, founder and former spokesperson of the British Centre for
Science Education (BCSE). When I wrote an open letter to American scientific organizations, criticizing their claims that professing nonbelief is inimical to the public acceptance of evolution, Stanyard commented on my
Web site:

> We have a political battle, to keep the creationists out of state-funded
> schools. It requires our very limited resources to be tightly focused on
> what is a single issue matter.
>
> Moreover, it will fail if it involves a general attack on religion because:
>
> 1. You'll lose a pile of allies.
> 2. The message immediately becomes confused and will be ignored.
> 3. It will immediately lead to a huge over stretching of resources.

If you want creationism out of schools, it's not an intellectual battle.
It's politics and you have to play politics to win. That includes forming
alliances with whom you might find distasteful and keeping your dis-
tance from many you might agree with.

The Varieties of Accommodationism

The brands of accommodationism fall into just a few categories, and I'll describe them in increasing order of intellectual rigor.

Logical Compatibility

This argument is used but rarely, for its claim is merely that there is no *logical* reason why religion and science are incompatible. It could be true, for instance, that deities exist who have never interfered with the workings of the universe. But pure deism is a rare brand of faith. Alternatively, there could be religions that alter their doctrines every time one becomes incompatible with a new finding of science, or with reason itself. Such beliefs, though compatible with science, are almost nonexistent as well. Ultraliberal forms of Christianity, like those that accept a personal God but deny miracles such as the Resurrection, may accept science but nevertheless still defy reason by making unsubstantiated claims.

Mental Compatibility

This is the most common—though hardly the most convincing—argument for accommodationism. It runs like this: "Many scientists are religious, and many religious people embrace science, so the two *must* be compatible." It is what Stephen Jay Gould had in mind when he said, "Unless at least half of my colleagues are inconsiderate dunces there can be—on the most raw and direct empirical grounds—no conflict between science and religion."

People making this claim often point to famous religious scientists of the past, like Isaac Newton, Robert Boyle, and J. J. Thomson. But of course in the early days of science *everyone* was religious, so this hardly counts as evidence of compatibility. Because publicly professed belief was ubiquitous, religion could be touted as compatible with all human endeavors.

What we must consider more seriously are the modern scientists who publicly avow their faith, for scientists have little to lose by professing atheism. Such "scientists of faith" include Francis Collins, a geneticist and an evangelical

Christian, the Anglican paleontologist Simon Conway Morris, and the Catholic cell biologist Kenneth Miller. Even Indian scientists, before they launch spacecraft, visit Hindu temples to ask their deities for blessings. In America, religious scientists aren't rare: a survey by the sociologist Elaine Ecklund showed that 23 percent of U.S. scientists believe in God with varying degrees of confidence, even though that's only one-fourth the proportion of believers among Americans as a whole.

Does this class of modern believers, then, demonstrate that science is compatible with religion? Well, that would be a very odd kind of compatibility, for it simply construes "compatibility" as the ability of two divergent worldviews to be simultaneously held in one person's mind. That says nothing about whether those views, or the methodologies they employ, are "accordant, consistent, congruous, or agreeable." This form of accommodationism confuses *coexistence* with *compatibility*.

And if religion and science are compatible in this way, so are marriage and adultery. After all, many married people are unrepentant adulterers. Astrology and science also become compatible, because many science-friendly people still consult their horoscopes. Indeed, because some scientists—often chemists or engineers—believe that the Earth is less than ten thousand years old, and that God created all species simultaneously, we might even say that science and *creationism* are compatible! We all know people who hold incompatible views, whether or not that causes them distress.

Syncretism

According to the *Oxford English Dictionary*, syncretism is the "attempted union or reconciliation of diverse or opposite tenets or practices, esp. in philosophy or religion." When discussing science and religion, syncretists claim that they are two sides of a single practice: finding truth. They're claimed to be harmonious in diverse ways, including seeing the cosmos and its laws as a religion ("pantheism"), holding that science and religion *cannot* conflict because they're both God-given ways of approaching truth, and claiming that the truths of science are already embodied in ancient scripture.

Syncretism, then, makes science and religion compatible by redefining one so that it includes the other. We may argue, for instance, that "God" is simply the name we give to the order and harmony of the universe, the laws of physics and chemistry, the beauty of nature, and so forth. This is the naturalistic pantheism of Spinoza, whose most famous recent advocate was Albert Einstein, often—and wrongly—described as accepting a personal God. One of Einstein's quotes is often used to show this:

> The most beautiful and deepest experience a man can have is the sense of the mysterious. It is the underlying principle of religion as well as of all serious endeavour in art and science. He who never had this experience seems to me, if not dead, then at least blind. To sense that behind anything that can be experienced there is a something that our minds cannot grasp, whose beauty and sublimity reaches us only indirectly: this is religiousness. In this sense I am religious. To me it suffices to wonder at these secrets and to attempt humbly to grasp with my mind a mere image of the lofty structure of all there is.

But this is clearly a paean not to the Abrahamic God, but to the mysteries of the universe. Although Einstein isn't the final authority on the harmony between science and faith, as he grew older his spirituality became increasingly synonymous with the laws of nature, and at odds with the religions of his American countrymen. In his biography of Einstein, Walter Isaacson recounts how Herbert Goldstein, an Orthodox Jewish rabbi in New York, sent Einstein a telegram asking directly, "Do you believe in God?" Einstein answered, "I believe in Spinoza's God, who reveals himself in the lawful harmony of all that exists, but not in a God who concerns himself with the fate and the doings of mankind." In other words, Einstein was at best a pantheist. The reason why accommodationists are so obsessed with Einstein's view of religion is that he's often viewed as the smartest man in history, so his approbation of religion would give faith a special imprimatur.

The big problem with pantheism, in which science does not marry religion so much as digest it, is that it leaves out God completely—or at least a monotheistic God who has an interest in the universe. Such a god is unacceptable to most religious people. As we've learned, more than 65 percent of Americans believe in a personal God who interacts with the world, as well as in the divinity of Jesus, heaven, and miracles. In his popular book *Finding Darwin's God*, Kenneth Miller attacked pantheism because it "dilutes religion to the point of meaninglessness." He added, "Such 'Gods' aren't God at all—they are just clever and disingenuous restatements of empirical science contrived to wrap an appearance of religion around them, and they have neither religious nor scientific significance." Most believers would probably agree.

Another syncretic argument is to equate "spirituality" with religion, disregarding the diverse forms of spirituality, many having nothing to do with the supernatural. Interviewed by *National Geographic* at his study site in Egypt, Philip Gingerich, a renowned paleontologist who has done seminal work on the evolution of whales, synonymized religion and spirituality:

> Gingerich is still baffled by the conflict that many people feel between religion and science. On my last night in Wadi Hitan, we walked a little distance from camp under a dome of brilliant stars. "I guess I've never been particularly devout," he said. "But I consider my work to be very spiritual. Just imagining those whales swimming around here, how they lived and died, how the world has changed—all this puts you in touch with something much bigger than yourself, your community, or your everyday existence." He spread his arms, taking in the dark horizon and the desert with its sandstone wind sculptures and its countless silent whales. "There's room here for all the religion you could possibly want."

Gingerich's "spirituality," which he clearly sees as religious, nevertheless approaches the kind of emotion described by Albert Einstein. And there are few scientists who haven't felt that way about either their work or the amazing discoveries constantly falling into the hopper of science. Further, despite the stereotype of scientists as cold automatons, immune to beauty and devoid of wonder, we're people first. We love the arts (I'd maintain that scientists appreciate the arts more than humanities scholars appreciate science); we have the same emotions as everyone else (as well as a special kind: the wonder we feel when discovering something that no one ever knew before); and sometimes we feel individually insignificant but connected to the larger universe we study.

If emotion, awe, wonder, and yearning are considered "spirituality," then call me spiritual, for I often feel the same "frisson in the breast" described by Richard Dawkins, a die-hard atheist, as his own form of spirituality. But emotionality isn't the same as religious belief in the divine or the supernatural, and it's not helpful for scientists like Gingerich to conflate them.

Yet the effort goes on. Elaine Ecklund, whose work on sociology is funded by the Templeton Foundation, has devoted much of her career to showing that scientists are more religious than everyone thinks. When she and her collaborator Elizabeth Long surveyed spirituality among scientists, they concluded, "Our results show unexpectedly that the majority of scientists at top research universities consider themselves 'spiritual.'" In reality, Ecklund and Long's "majority" was only 26 percent of all scientists—barely a quarter! The authors go on to admit, "Our results show that scientists hold religion and spirituality as being qualitatively different kinds of constructs," adding that "in contrast to their views on spirituality, what scientists in this group specifically dislike about religion is the sense of faith that they think often leads religious people to believe without evidence." The difference between a "religious" and a "spiritual" scientist couldn't be clearer, but Ecklund still uses this kind of data to argue for harmony between science and religion.

Other syncretists argue that it's *impossible* for science to contradict religion because science, devoted to understanding how God's creation works, *must* comport with religious belief. As Pope John Paul II put it in 1996, "truth cannot contradict truth." The theologian Stephen J. Pope from Boston College explained further: "God is the source of both reason and revelation, and truth from one source cannot contradict truth from the other. Disagreements in science and religion are capable of reconciliation because these sources are two valid but distinct modes of apprehending what is true."

But such claims look silly when the two areas provide "answers" that are irreconcilable—the "conflict of outcomes" described in the last chapter. Perhaps the best example of this forced harmony is "scientific creationism," a movement that began in America in the 1960s and died out about twenty years later. After American courts rejected the teaching of biblical creationism in public school science classes on constitutional grounds (creationism was seen as a form of religion, violating our legal separation between church

and state), creationists regrouped under the rubric of "scientific creationism," claiming that the findings of science were perfectly reconcilable with
the Bible. They could then argue that teaching biblical ideas wasn't religious
at all, but simply science.

That too failed, for the reconciliation is spurious. To harmonize the fossil record with the story of Noah's flood, for instance, scientific creationism
proposed the ludicrous theory of "hydrodynamic sorting," arguing that a
sudden worldwide flood would in fact yield precisely the fossil record we
see. Marine invertebrates, living on the seafloor, would naturally be the first
to be covered with sediment when the waters began to rise and roil. They
would therefore show up at the bottom of the geological record, the part
that scientists consider the oldest. The fishes would follow, settling atop the
invertebrates, and then, in order, we'd see amphibians (who live close to
water), reptiles, and then mammals, who, being smarter and more agile,
would be able to flee the rising waters. And humans, the smartest and
most resourceful of all creatures, could climb quite high before they were
inundated, accounting for our appearance as fossils in the topmost geological layer.

As a young assistant professor, I taught a course called "Evolution vs. Scientific Creationism"—perhaps the most fun I've ever had as a teacher. On
Mondays I'd lecture as an evolutionary biologist, and on Wednesdays as a
creationist, refuting what I had said on Monday. (I was already quite familiar
with the claims of creationists, and could easily talk like one.) The students, of
course, became deeply confused. But on Fridays we'd have a discussion and
sort out the competing claims. And it was when we came to the "hydrodynamic sorting" hypothesis that the students realized that biblical "truth" simply couldn't be harmonized with scientific truth. Why weren't some unlucky
humans, perhaps confined to beds or wheelchairs, buried in sediments alongside fossil amphibians? Why didn't some seagoing mammals, like whales,
sleep with the fossil fishes, instead of appearing later alongside the mammals?
And why did the flying pterodactyls get inundated so much earlier than modern birds, when both could fly to the mountaintops? Such is the debacle that
results from claiming that "truth cannot contradict truth." Eventually, scientific creationism went the way of its literalist ancestor, branded by the courts
as simply fundamentalism tricked out as science.

But this strategy is still alive. As we'll see in the next chapter, "natural theology," the idea that some facts of science support the existence of God— and in fact can't be explained *except* by God—is alive and well among even liberal theologians.

Muslim accommodationists, who, like most Muslims, take the Quran literally, have their own form of scientific creationism, asserting that the book is not only scientifically accurate on all issues, but actually anticipated every finding of modern science. The results are both pathetic and amusing. The Turkish physician Halûk Nurbaki, for instance, collected fifty verses from the Quran, striving mightily to show that they predicted the discovery of gravity, the atomic nucleus, the Big Bang, and quantum mechanics. He translated one such verse as "The fire you kindle arises from green trees." Nurbaki sees this as a divine indication of the oxygen produced by plants and consumed by fire, adding, "It was impossible 14 centuries ago for unbelievers to understand the stupendous biological secret this verse contains, for the inside story of combustion was not known." All this shows is how far some people can twist scripture to make their faith comport with science. (The one exception for Muslims is human evolution: while many have no problem with evolution itself, they nearly all agree with the Quran that our species is unique, created instantly by Allah from a lump of mud. And nearly all Muslim science classes exempt humans from the evolutionary process.)

Because accommodationism is largely a Western enterprise, it's harder to find works in English that reconcile Eastern faiths with science. But one strain of Hindutva, the growing Hindu nationalist movement, apparently does for the Vedas what Nurbaki does for the Quran, forcing science into the Procrustean bed of scripture. As the Indian historian and philosopher Meera Nanda notes, this current of thought "simply grabs whatever theory of physics or biology may be popular with Western scientists at any given time, and claims that Hindu ideas are 'like that,' or 'mean the same' and 'therefore' are perfectly modern and rational."

What about Buddhists? That "faith," of course, comprises many sects, some more philosophical than religious, but the most famous statement on Buddhist accommodationism comes from Tenzin Gyatso, the current Dalai Lama. Fascinated by science from his youth, Gyatso wrote an entire book trying to harmonize science and Buddhism, *The Universe in a Single Atom*. It contains a statement often quoted to show the primacy of fact over faith in Buddhist teaching: "If scientific analysis were conclusively to demonstrate certain claims in Buddhism to be false, then we must accept the findings of science and abandon those claims." Yet Gyatso nevertheless accepts at least two supernatural claims, reincarnation and the "law of karma," and criticizes the theory of evolution along creationist lines, arguing that mutations aren't random and that the notion of "survival of the fittest" is a tautology (it isn't). Buddhism is considered one of the "nonliteralist" faiths, but, like all faiths, its literalism about some beliefs makes it incompatible with science.

The NOMA Gambit

The most famous attempt to reconcile science and religion, at least in recent years, was made by the paleontologist Stephen Jay Gould. In his 1999 book *Rocks of Ages: Science and Religion in the Fullness of Life,* Gould argued that the compatibility of science and religion rests on understanding that their aims are completely separate. Science, he said, is the endeavor to find out about the natural world, while religion deals solely with issues of meaning, purpose, and morals. The two disciplines thus constitute "non-overlapping magisteria," for which Gould coined the acronym NOMA. To Gould, this disjunction creates a kind of harmony: dealing with the human condition, he argued, requires both physical and metaphysical inquiry. The NOMA argument has been co-opted by many scientific organizations eager to show that they don't step on religion's toes.

Gould wasn't the first to float this idea: both theologians and philosophers previously made similar claims. The mathematician Alfred North Whitehead, for instance, anticipated Gould in 1925:

> Remember the widely different aspects of events which are dealt with in science and in religion respectively. Science is concerned with the general conditions which are observed to regulate physical phenomena; whereas religion is wholly wrapped up in the contemplation of moral and aesthetic values. On the one side there is the law of gravitation, and on the other the contemplation of the beauty of holiness. What one side sees, the other misses; and vice versa.

Gould's contribution was not only to formalize this argument in an entire book, but also to promote it as a principle of sound intellectual behavior. Its popularity—for the idea is neither new nor profound—undoubtedly reflects Gould's compelling prose, the "let's all get along" tone of the book, and the fact that an argument "that grants dignity and distinction to *each* subject" was being made by a famous and popular scientist who was also an outspoken atheist.

Unfortunately, Gould's attempt fails on two counts: it requires the homeopathic dilution of religion into a humanistic philosophy devoid of supernatural claims, and it gives to religion sole authority over moral and philosophical issues that have nevertheless had a long secular history. Because NOMA is perhaps the most common argument for the compatibility of science and faith, it bears some examination.

Gould began his argument by observing that both science and religion have sometimes transgressed their proper boundaries, with religion in effect making scientifically testable statements about nature, and scientists inferring ethical or social principles from nature. The most obvious example of the former is American creationism; of the latter are early attempts to justify racism and capitalism by appealing to the theory of evolution. Using examples drawn from the work of Darwin, Galileo, Cardinal Newman, and other scientists and theologians, Gould showed that these territorial violations have occurred throughout history. NOMA, he argued, will prevent them from recurring if we simply stick to the following principles:

> Science tries to document the factual character of the natural world, and to develop theories that coordinate and explain these facts. Religion, on the other hand, operates in the equally important, but utterly different, realm of human purposes, meanings, and values—subjects that the factual domain of science might illuminate, but can never resolve.

Gould thus granted these magisteria "equal importance," calling for thoughtful dialogue between religion and science—not to unite them, but to encourage greater harmony and mutual understanding.

The problem is that while NOMA appeals as a utopian vision, Gould saw it as more than just a pleasing platitude, for he urged that we must realize his vision by *structuring* science and religion in a way that would allow their peaceful coexistence. He thus saw NOMA as "the potential harmony through difference of science and religion, both properly conceived and limited."

The word "properly" is the red flag here. Imagining "proper" science is easy—the vast majority of scientists are happy to pursue their calling as an entirely naturalistic enterprise. But what is "proper" religion? It was, to Gould, religion that does not overlap with science.

And that's the rub, for real religion is frequently and stubbornly *improper*. As we've seen, many people's religions, by making factual claims about the world, bring them into Gould's territory of science. As always, evolution is the most prominent example. It's not only fundamentalists who subscribe to unscientific creationist narratives, but also many mainstream Protestants and Catholics, Mormons, Jehovah's Witnesses, Orthodox Jews, Native Americans, Scientologists, Muslims, and Hindus. But ideas about the origin of humans and other species aren't the only religious violations of NOMA. Christian Scientists entertain a spiritual theory of disease, and some Hindus believe that disability is a sign of past spiritual transgression. Most Abrahamic religions accept the existence of souls that distinguish humans from other species. It's simply undeniable that religions worldwide often stray into scientific territory, sometimes with tragic results. How many have died, even in the last few decades, because an infection is regarded as simply spiritual malaise?

To deal with this difficulty, Gould apparently construed "religion" as the pronouncements of liberal Western theologians, many of them agnostics in all but name. But of course there is far more to religion than the opinions of scholars. Religion encompasses beliefs that help people make sense of personal reality, even when those beliefs overlap with science. By casting himself as the arbiter of "proper" religion, Gould simply redefined terms to satisfy his utopian vision. Thus NOMA underwent a second metamorphosis from an achievable utopia to *an actual description of reality*. That is, to Gould the distressing clashes between faith and science *by definition* did not involve "real" religion. This turned NOMA into an exercise in tautology, allowing him to simply dismiss religions that make claims about reality:

Religion just can't be equated with Genesis literalism, the miracle of the liquefying blood of Saint Januarius . . . or the Bible codes of kabbalah and modern media hype. If these colleagues wish to fight superstition, irrationalism, philistinism, ignorance, dogma, and a host of other insults to the human intellect (often politically converted into dangerous tools of murder and oppression as well), then God bless them—but don't call this enemy "religion."

But what else can we call it? Many religious people would be affronted to learn that NOMA requires them to abandon essential parts of their faith. Nevertheless, that was apparently Gould's prescription. He denied to religion, for instance, a reliance on miracles, arguing that the first commandment for NOMA is "Thou shalt not mix the magisteria by claiming that God directly ordains important events in the history of nature by special interference knowable only through revelation and not accessible to science." But of course this rejects the central claim of Christianity—the Resurrection—as well as the Catholic and literalist beliefs in a historical Adam and Eve.

And what about the most obvious violation of NOMA: the many forms of religion that accept creationism? To save his argument, Gould contended that creationism is neither proper religion nor even an *outgrowth* of religion:

> In other words, our struggle with creationism is political and specific, not religious at all, and not even intellectual in any genuine sense. . . . Creationists do not represent the magisterium of religion. They zealously promote a particular theological doctrine—an intellectually marginal and demographically minority view of religion that they long to impose on the entire world.

Sadly, this argument is nonsense. Anyone who has battled creationism or its city cousin "intelligent design" realizes that these are purely religious phenomena, born of the conflict between evolutionary biology and scripture. Scratch a creationist and at least 99 percent of the time you'll find a religionist.

And it's not just biblical literalists who decry evolution. Many adherents to more moderate faiths, including Catholicism, reject evolution because of its unsavory implications for human morality and uniqueness. Recall that 42 percent of Americans are creationists with respect to humans, but biblical fundamentalists are far fewer, and 82 percent of Americans think that some form of creationism should be taught in public schools, either by itself or alongside conventional evolutionary theory. Nor is creationism a "demographically minority view of religion," at least in America, for, as we've seen, 73 percent of all Americans violate NOMA by seeing at least some acts of God as responsible for living species—hardly a demographic minority.

Finally, Gould's designation of religion as the preserve of morals, meaning, and purpose is both disingenuous and historically inaccurate. For one thing, it ignores a centuries-long debate about the source of ethical belief. Does religion directly create moral views, or does it only codify and reinforce morality that flows from secular springs?

Gould sensed this difficulty but again finessed it by redefinition: all ethics, he claimed, is really religion in disguise. Trying to distinguish the two is, he said, to simply "quibble about the labels," and so he chose to "construe as fundamentally religious (literally, binding us together) all moral discourse on principles that might activate the ideal of universal fellowship of people." But serious scholarly discussion of ethics really began as a secular endeavor in ancient Greece, continued in a nonreligious vein through philosophers like Kant and Mill, and in our day persists among atheist philosophers like Peter Singer and Anthony Grayling. The majority of modern ethical philosophers are in fact atheists. By eliminating the empirical claims of religion but stretching it to cover ethics and "meaning," Gould simultaneously shrank and expanded religion.

But what about the other side—the NOMA violations of scientists? Yes, we've had them; biology textbooks of the early twentieth century, for instance, contain chapters on eugenics that repel us today. But these days it's quite rare to find scientists drawing moral lessons from their own work, much less trying to impose them on society. Most scientists have become quite cautious about overstepping their magisterium, and the crimes that Gould imputes to such boundary violations, including lynchings, the horrors of both world wars, and the bombing of Hiroshima, have little to do with science itself and more with the appropriation of technology by people lacking foresight or morals. (As I'll argue later, this contrasts with religion, whose malfeasance is a direct by-product of the moral codes inherent in most faiths.) And most of science's "violations" were history by the time Gould wrote his book, in contrast to the many empirical claims still made by religion.

By and large, scientists now avoid the "naturalistic fallacy"—the error of drawing moral lessons from observations of nature. They've therefore adopted the NOMA principle far more readily than have philosophers or theologians. Ethical philosophers, particularly the nonbelievers, are rightly peeved to learn that their labors now fall under the rubric of "religion." But believers are even more upset. Out of the thousands of religious sects on this planet, only a handful lack adherents or dogmas that make empirical claims about the cosmos. A religion whose god does not interact with the world—that is, a religion considered "proper" by Gould's lights—is a religion whose god is absent. Honest believers admit this. One of them is Ian Hutchinson, a Christian physicist:

> But the religion [Gould] is making room for is empty of any claims to historical or scientific fact, doctrinal authority, and supernatural experience. Such a religion, whatever be its attractions to the liberal scientistic mind, could never be Christianity, or for that matter, Judaism or Islam.

The theologian John Haught agrees:

> [A] closer look at Gould's writings about science and religion will show that he could reconcile them only by understanding religion in a way that most religious people themselves cannot countenance. Contrary to the nearly universal religious sense that religion puts us in touch with the true depths of the real, Gould denied by implication that religion can ever give us anything like reliable knowledge of *what is*. That is the job of science alone. . . . Still, Gould could not espouse the idea that religion in any sense gives us truth.

In the end, NOMA is simply an unsatisfying quarrel about labels that, unless you profess a watery deism, cannot reconcile science and religion. As Isaiah realized when prophesying harmony among the beasts, it takes a miracle to reconcile the irreconcilable: "And the lion shall eat straw like the ox."

Science Versus the Supernatural

Implicit in the NOMA gambit is the claim that science deals only with questions involving natural phenomena, while questions about the supernatural fall in the bailiwick of religion. You'll often see this claim made by scientific organizations trying to avoid alienating believers. Here's such a claim from a very prestigious group of scientists, the National Academies:

> Because they are not a part of nature, supernatural entities cannot be investigated by science. In this sense, science and religion are separate and address aspects of human understanding in different ways. Attempts to pit science and religion against each other create controversy where none needs to exist.

The National Science Teachers Association made a similar proclamation:

> Science is a method of testing natural explanations for natural objects and events. Phenomena that can be observed or measured are amenable to scientific investigation. Science also is based on the observation that the universe operates according to regularities that can be discovered and understood through scientific investigations. Explanations that are not consistent with empirical evidence or that cannot be tested empirically are not a part of science. As a result, explanations of natural phenomena that are not derived from evidence but from myths, personal beliefs, religious values, philosophical axioms, and superstitions are not scientific. Furthermore, because science is limited to explaining natural phenomena through testing based on the use of empirical evidence, it cannot provide religious or ultimate explanations.

Many liberal churches have issued similar statements. By reassuring people that science has nothing to say about their faith, such words are supposed to turn creationists into supporters of evolution. Sadly, there's no evidence that this has worked.

The main problem, however, is that these statements are flatly wrong. Science in fact has a lot to say about the supernatural. It can and has tested it, and so far has found no evidence for it.

But let's back up, for such a statement demands that we clarify what we mean by "supernatural." One way is to simply claim that the supernatural is "the realm of phenomena that can't be studied by scientific methods." This is the sense used by the philosopher of science Robert Pennock when arguing that "[if] we could apply natural knowledge to understand supernatural powers, then, by definition, they would not be supernatural." Such a definition makes the statements by the National Academies and the National Science Teachers Association true, but only as tautologies.

As we've already seen, the *Oxford English Dictionary* defines "supernatural" as "belonging to a realm or system that transcends nature, as that of divine, magical, or ghostly beings; attributed to or thought to reveal some force beyond scientific understanding or the laws of nature; occult, paranormal." In other words, the supernatural includes those phenomena that violate the known laws of nature. What's important to realize is that this definition does not make the supernatural off-limits to science. Indeed, over its history science has repeatedly investigated supernatural claims and, in principle, could find strong evidence for them. But that evidence hasn't appeared.

Note too that the supernatural includes not only divine phenomena (which I'd characterize as things caused by beings having mind but no substance), but also the *paranormal:* phenomena like alchemy, homeopathy, ESP, telekinesis, ghosts, astrology, Buddhist karma, and so on. All of these involve breaking the known laws of nature. Because we surely don't know all the laws of nature, one must consider that something that appears "supernatural" to science may lose that status with further study. As I describe below, there are certain observations that would convince me of the truth of some religions, but that truth might ultimately be due to misunderstanding—a giant magic trick played by space aliens, for instance. After all, in science all conclusions about the universe are provisional. But it would be a mistake for scientists to completely rule out a priori any truly supernatural phenomena, religious or otherwise.

Nearly all religions make empirical claims about how God interacts with the world, although some of them are hard or impossible to test. Contra Gould, this means that most religions overstep their NOMA boundaries. I've already mentioned some religious claims about reality, but let's look at how science might test whether phenomena are supernatural acts of a god, as well as the existence of a god itself. The philosophers Yonatan Fishman and Maarten Boudry gave seven such tests, describing which outcomes would give evidence for a god or other supernatural and paranormal phenomena. (There's no real difference between "supernatural" and "paranormal" phenomena: both involve violating the known laws of nature, though the former term usually refers to divine intervention and the latter to "nonreligious" phenomena like ESP and clairvoyance.)

1. Intercessory prayer can heal the sick or re-grow amputated limbs.
2. Only Catholic intercessory prayers are effective.
3. Anyone who speaks the Prophet Mohammed's name in vain is immediately struck down by lightning, and those who pray to Allah five times a day are free from disease and misfortune.
4. Gross inconsistencies found in the fossil record and independent dating techniques suggest that the earth is less than 10,000 years old—thereby confirming the biblical account and casting doubt upon Darwinian evolution and contemporary scientific accounts of geology and cosmology.
5. Specific information or prophecies claimed to be acquired during near-death experiences or via divine revelation are later confirmed—assuming that conventional means of obtaining this information have been effectively ruled out.
6. Scientific demonstration of extra-sensory perception or other paranormal phenomena (e.g., psychics routinely win the lottery).
7. Mental faculties persist despite destruction of the physical brain, thus supporting the existence of a soul that can survive bodily death.

Now, we already have anecdotal evidence against nearly all of these claims, and more systematic evidence against ESP and the efficacy of prayer. The most relevant studies for our purposes are those of prayer.

It's surprising how often Americans pray, and how much confidence they have that it works. Eighty-eight percent of Americans pray to God, 76 percent say prayer is an important part of their daily lives, and 83 percent believe that there is a God who answers prayers. And those prayers aren't just meditative exercises. More than 50 percent of Christians, Jews, and Muslims pray for their health and safety, good relationships with others, and help with mental or physical illnesses.

Chapter 5 describes the dreadful outcomes of using prayer instead of medicine for healing, but many Americans use prayer as a *supplement* to doctors. Over 35 percent of Americans pray for their health in a given year, and 24 percent ask others to do so. Clearly these people believe that prayers, both their own and others', can work.

If you believe prayer works, and isn't just a way of having a chat with God, then that belief can be tested. In fact, such a test was first conducted in 1872 by the geneticist and statistician Francis Galton, Charles Darwin's half cousin. Galton figured that among all British males who lived at least thirty years, those who were prayed for most often would be the regents ("God save the King"). If that were so, you'd expect that kings would, on average, live longer than other males, including the aristocracy, clergy, artists, tradesmen, and doctors themselves. (Kings also have the advantage of better food and medical care.) Contrary to Galton's hypothesis, though, ninety-seven sovereigns examined had the *shortest* longevity of all the classes tested: sixty-four years as opposed to averages between sixty-seven and seventy years. But while Galton fobbed off petitionary prayer as a remnant of ancient superstition, he hedged by suggesting that it still might be useful for communing with any gods, relieving stress, and bringing strength. He apparently believed in belief.

One can, of course, dismiss this study as a lighthearted investigation of a passing idea (Galton was prone to such spur-of-the-moment statistical tests), but there are more modern and scientifically controlled studies of the effects of intercessory prayer on healing. Three of the best involved the effect of prayer on recovery after hospitalization for heart problems, after cardiac catheterization or angioplasty, and the effect of "distant healing" of patients after breast-reconstruction surgery. All three had proper controls: that is, some patients weren't prayed for, and none *knew* whether they were prayed for. The results were uniformly negative: there were no positive effects of prayer on healing. A somewhat smaller study of healing after breast surgery also showed no effect of prayer, and a combination of prayer and other distant-healing methods had no effect on the medical and psychological condition of patients carrying HIV.

Theists' typical response to these failures is to say either "God won't let himself be tested" or "That's not what prayer is about: it's simply a way to converse with God." But you can bet that had these studies shown a large positive effect, the religious would be noisily flaunting this as evidence for God. The confirmation bias shown by accepting positive results but explaining away negative ones is an important difference between science and religion.

There's no substantive difference between the paranormal and the supernatural, and demonstrations of paranormal phenomena have also failed in medicine—including nonreligious methods of "cure" such as therapeutic touch, inhaling flower scents, and using magnets—as well as in other areas like ESP, telekinesis, and past-life regression. One can envision many other tests of religious claims. Does rain dancing help Native Americans relieve drought? Can God affect evolution by raising the probability of an adaptive mutation when conditions change?

So what would convince a skeptic like me of a miracle—a phenomenon that violated the laws of nature? Several of Fishman and Boudry's examples from the list above would suffice. And because humans, unlike salamanders, don't have the ability to regenerate lost limbs, if a religious healer could repeatedly regrow missing limbs by saying prayers over the afflicted, and this was documented with reliable evidence and testimony by multiple doctors, I would consider that a miracle, and perhaps evidence for God. But it hasn't escaped people's notice that "miracle" healings are always of the kind, like the disappearance of tumors, that can happen naturally, even without prayer. The Vatican itself, which requires a miracle to beatify someone, and two miracles to make that person a saint, is none too scrupulous about the medical evidence needed to elevate someone to the pantheon. The "miracle" that clinched the beatification of Mother Teresa, for instance, was the supposed disappearance of ovarian cancer in Monica Besra, an Indian woman who reported she was cured after looking at a picture of the nun. It turns out, though, that her tumor wasn't cancerous but tubercular, and, more important, she'd received conventional medical treatment in a hospital, with her doctor (who wasn't interviewed by the Vatican) taking credit for the cure.

More convincing forms of healing are simply never seen. Anatole France brought this up in his book *Le Jardin d'Épicure:*

> When I was at Lourdes in August, I visited the grotto where innumerable crutches had been put on display as a sign of miraculous healing. My companion pointed out these trophies of illness and whispered in my ear:
> "One single wooden leg would have been much more convincing."

Indeed. The question "Why won't God heal amputees?" is almost a cliché of atheism, but isn't it reasonable to ask why wooden legs and glass eyes aren't on exhibit at Lourdes? France had a response:

> That seems sensible, but, philosophically speaking, the wooden leg has no more value than a crutch. If an observer with true scientific spirit witnessed the regrowing of a man's severed leg after immersion in a sacred pool or the like, he would not say "Voilà—a miracle!" Rather, he would say, "A single observation like this would lead us to believe only that circumstances we don't fully understand could regrow the leg tissues of a human—just like they regrow the claws of lobsters or the tails of lizards, but much faster."

Here France rejects the supernatural in favor of natural laws that we haven't yet discovered. Such healings, for example, could be the work of altruistic space aliens with advanced abilities to regrow tissue. But it doesn't matter. If we consider the regeneration of limbs or eyes not as absolute evidence for God, but—as a scientist would—*provisional* evidence, then it points us toward the divine. And if these miracles occur repeatedly, are documented carefully, and occur only under religious circumstances, then the evidence for a supernatural power grows stronger.

In his book *The Varieties of Scientific Experience*—deliberately named to mimic William James's classic study of religion—Carl Sagan describes how ancient scripture *could* have given us scientific evidence for God. It could, for instance, have presented information not known to humans when the sacred texts were written. These include statements like, "Thou shalt not travel faster than light" or "Two strands entwined is the secret of life." God could also have made his presence known by engraving the Ten Commandments in large letters on the Moon. Unless defined tautologically, then, the supernatural is either in principle or in practice within the realm of science. And when we consider all the failures to find it—the lack of accurate predictions in scripture, the failure of science to confirm testable religious claims, the failure of a god to make its presence unimpeachably known—we find a big hole: the absence of evidence when the evidence *should be there*. Our rational response should be to tentatively reject the existence of any supernatural beings or powers.

Evidence for the supernatural, of course, is not evidence for a god or, especially, for the tenets of a particular religion. That requires other information. But some nonbelievers reject the possibility of *any* evidence for gods, claiming that the concept of a god itself is so nebulous, so incoherent, that there could never be evidence to support one. I disagree, and I think most scientists could think of some observations that would convince them of the existence of God. Even Darwin himself had some ideas, which he mentioned in a letter to the American botanist Asa Gray in 1861:

> Your question what would convince me of Design is a poser. If I saw an angel come down to teach us good, & I was convinced, from others seeing him, that I was not mad, I shd. believe in design.—If I could be convinced thoroughily [*sic*] that life & mind was in an unknown way a function of other imponderable forces, I shd. be convinced.—If man was made of brass or iron & no way connected with any other organism which had ever lived, I shd. perhaps be convinced. But this is childish writing.

Well, perhaps not so childish, for it tells us that Darwin, like a good scientist, was open to evidence for "Design," by which he surely meant "God." I too could be convinced of the Christian God. The following (and admittedly contorted) scenario would give me tentative evidence for Christianity. Suppose that a bright light appeared in the heavens, and, supported by wingèd angels, a being clad in a white robe and sandals descended onto my campus from the sky, accompanied by a pack of apostles bearing the names given in the Bible. Loud heavenly music, with the blaring of trumpets, is heard everywhere. The robed being, who identifies himself as Jesus, repairs to the nearby university hospital and instantly heals many severely afflicted people, including amputees. After a while Jesus and his minions, supported by angels, ascend back into the sky with another chorus of music. The heavens swiftly darken, there are flashes of lightning and peals of thunder, and in an instant the skies clear.

If this were all witnessed by others and documented by video, and if the healings were unexplainable but supported by testimony from multiple doctors, and if all the apparitions and events conformed to Christian theology—then I'd have to start thinking seriously about the truth of Christianity. Perhaps such eyewitness evidence isn't even necessary. If, as Sagan suggested, the New Testament contained unequivocal information about DNA, evolution, quantum mechanics, or other scientific phenomena that couldn't have been known to its authors, it would be hard not to accept some divine inspiration.

Perhaps other scientists would call me credulous. My scenario about a visiting Jesus could, they say, be a gigantic con game played by aliens with the technology to pull off such a stunt. (Curiously, those who make such arguments never extend them to their logical conclusion, that all life on Earth could be merely a Matrix-like computer simulation run by aliens.) After all, the science-fiction writer Arthur C. Clarke's "third law" was "Any sufficiently advanced technology is indistinguishable from magic." But I think you can substitute "God" for "magic." And this is why my acceptance of God would be provisional, subject to revocation if a naturalistic explanation arose later. Extraordinary claims require extraordinary evidence, but we can never say that such evidence is impossible.

Now turn the question around: ask religious people what evidence it would take to make them abandon their faith. While some will actually give thoughtful responses, what you'll hear most often is the answer of Karl Giberson cited in the previous chapter: *no* data could dispel his belief in God.

He also gave some reasons for this stand, reasons that Christians don't often admit:

> As a purely practical matter, I have compelling reasons to believe in God. My parents are deeply committed Christians and would be devastated, were I to reject my faith. My wife and children believe in God, and we attend church together regularly. Most of my friends are believers. I have a job I love at a Christian college that would be forced to dismiss me if I were to reject the faith that underpins the mission of the college. Abandoning belief in God would be disruptive, sending my life completely off the rails.

This shows what we already know: belief may arise by indoctrination or authority, but is often maintained by social utility. But if no conceivable evidence can shake your faith in a theistic God, then you've deliberately removed yourself from rational discourse. In other words, your faith has trumped science.

What About Miracles?

Scientific analysis of miracles, at least those that happened in the distant past, suffers from two problems: determining whether they occurred at all, and determining whether they violated the laws of nature. If those miracles are supposed to be purposefully caused by a deity, that adds a third problem—or even a fourth if we want evidence that the miracles vindicate a specific faith like Christianity. Because miracles by definition can't be replicated, it's no coincidence that the pivotal doctrines of many religions now rest on ancient, one-off events like the dictation of the Quran by Allah and the Resurrection of Jesus. Must we then suspend judgment on such things? I think not. Let's take the Resurrection as an example.

The physicist Ian Hutchinson argues that the uniqueness of miracles makes them immune to science. If human levitation occurred repeatedly, he argues, science could test it, but "a religious faith that depended upon a belief that levitation was demonstrated on one particular occasion, or by one particular historic character, does not lend itself to such a scientific test. Science is powerless to bring unique events to the empirical bar."

But this can't be true, for historians have ways of confirming whether unique events are likely to have occurred. Those methods depend on multiple and independent corroboration of those events using details that coincide among different reporters, reliable documents that attest to those events, and accounts that are contemporaneous with the event. In this way we know, for example, that Julius Caesar was assassinated by a group of conspirators in the Roman Senate in 44 BCE, though we're not sure of his last words. As has been pointed out many times, the biblical account of the Crucifixion and Resurrection fails these elementary tests because the sources are not independent, none are by eyewitnesses, all contemporary writers outside of scripture fail to mention the event, and the details of the Resurrection and empty tomb—even among the Gospels and the letters of Paul—show serious discrepancies. Nor, despite ardent searching, have biblical archaeologists found such a tomb.

Theologians, of course, have their own arguments for why the Resurrection is true: Paul had a vision of the resurrected Christ; the empty tomb was found by women (bizarrely, some see this as "evidence" because a fictional Resurrection concocted in those sexist times would not involve the testimony of women); and although the scriptures and Paul's vision were not written down within Jesus's lifetime, they were described only a few decades later. But if you see that as convincing evidence, consider the "testimonies" that begin the Book of Mormon. Opening the book, you'll find two separate statements, signed by eleven named witnesses, all swearing they actually *saw* the golden plates given to Joseph Smith by the angel Moroni. Three of the witnesses—Oliver Cowdery, David Whitmer, and Martin Harris—add that an angel personally laid the plates before them. Unlike the story of Jesus, this is *actual eyewitness testimony*! Christians of other sects reject this testimony, but why then do they accept the tales about Jesus in the New Testament that are not only secondhand but produced by unknown writers? That's not a consistent way to deal with evidence.

The classic test for the truth of miracles is that of the Scottish philosopher David Hume. Foreshadowing Sagan's principle of "extraordinary evidence," Hume claimed that miracles were so extraordinary that to accept them, you would have to regard the suspension of nature's laws as *more* likely than any other explanation—including fraud or mistakes. When weighing evidence for nonmiraculous explanations, you should also consider if the witnesses stand to benefit from describing the miracle. Because we know that mistakes, fraud, and confirmation bias aren't that rare, to Hume they became the default explanation. Using this principle, if you reject the eyewitness testimony of eleven Mormons as fraud, error, or delusion, then you must also reject the Resurrection.

This is simply the scientific principle of parsimony: when you have several explanations for a phenomenon, it's usually (but not always) best to go with the one that has the fewest assumptions. And when miracles are sufficiently recent or sufficiently common to study scientifically, Hume's principle has held up. The Shroud of Turin, bearing the image of a half-naked man with wounds, has been considered for centuries as the burial shroud of Jesus, whose image was miraculously imprinted on the cloth. Although the Catholic Church does not give it official status as a genuine relic, it has been endorsed by several popes, including Pope Francis, in a way that implies that it might be real. Nevertheless, radiocarbon dating shows that the shroud was produced in medieval times, and its image of Jesus has been reproduced by an Italian chemist using materials available in medieval Europe.

In 2012, a statue of Jesus in Mumbai began oozing water from its feet. The "holy water," some of which was consumed, was seen as a miracle, and hundreds of Catholics flocked to worship the image. Unfortunately, the Indian skeptic Sanal Edamaruku discovered that the "miracle" was due to faulty plumbing: blocked drainage of a nearby toilet caused the statue to wick fecally contaminated water into the base, emerging at Jesus's feet. One would think that would settle the matter, but outraged believers caused Edamaruku to be indicted for violating Indian laws against hurting religious sentiments. He fled the country to avoid jail, a refugee from superstition. In an age of critical scrutiny and public media, the regular debunking of such miracles should give pause to those who see ancient miracles as genuine.

Hume's principle also promotes scientific reasoning in another way: look for alternative explanations. If you can think of a naturalistic, nondivine explanation for a "miracle," you should become agnostic about that miracle, and if you can't test it, then refuse to accept it. If there *are* such alternatives, the last thing you should do is make that miracle the pivot on which your whole faith turns.

There are, for instance, many alternative and nonmiraculous explanations for the story of Jesus's Resurrection. One was suggested by the philosopher Herman Philipse. It seems likely—for Jesus explicitly states this in three of the four Gospels—that his followers believed he would restore God's kingdom in their lifetime. Further, the apostles were told they'd receive ample rewards in their lifetime, including sitting on twelve thrones from which they'd judge the tribes of Israel. But, unexpectedly, Jesus was crucified, ending everyone's hope for glory. Philipse suggests that this produced painful cognitive dissonance, which in this case was resolved by "collaborative storytelling"—the same thing modern millennialists do when the world fails to end on schedule. The ever-disappointed millennialists usually agree on a story that somehow preserves their belief in the face of disconfirmation (for example, "We got the date wrong"). Philipse then suggests that in the case of the Jesus tale, the imminent arrival of God simply morphed into a promise of eternal life, a promise supported by pretending that their leader himself had been resurrected.

If you accept that an apocalyptic preacher named Jesus existed, who told his followers that God's kingdom was nigh, this story at least seems reasonable. After all, it's based on well-known features of human psychology: the behavior of disappointed cults and our well-known attempts to resolve cognitive dissonance. Like disillusioned millennialists, the early Christians could simply have revised their story. Is this really less credible than the idea that Jesus arose from the dead? Only if you have an a priori commitment to the myth.

It's no surprise then that the Jesus Seminar, a group of more than two hundred religious scholars charged with evaluating the historical truth of the words and deeds of Jesus, concluded that there was no credible evidence for either the Resurrection, the empty tomb, or Jesus's postmortem reappearance. They commented dryly, "The body of Jesus probably decayed as do all corpses." And they added a warning:

> The pre-eminent danger faced by Christian scholars assessing the gospels is the temptation to find what they would like to find. As a consequence, the inclination to fudge tends to be high—even among critical scholars—when working with traditions that have deep emotional roots

and whose critical evaluation has sweeping consequences for the religious community.

Of course, more conservative Christians have criticized this historical approach, even branding the work of the Jesus Seminar as heresy.

Does Hume's criterion mean, then, that we can never accept miracles? I don't think so, for Hume took it too far. *No* amount of evidence, it seems, could ever override his conviction that miracles were really the result of fraud, ignorance, or misrepresentation. Yet perhaps there are some events, though they're hard to imagine, when a divinely produced violation of nature's laws is more likely than human error or deception. It would be a close-minded scientist who would say that miracles are impossible in principle. But Hume was right about one thing: to have real confidence in a miracle, one needs evidence—massive, well-documented, and either replicated or independently corroborated evidence from multiple and reliable sources. No religious miracle even comes close to meeting those standards.

Three Test Cases

When science disproves religious beliefs that are negotiable—that is, parts of church doctrine that aren't critical parts of belief—the faithful are often happy to simply jettison them. Such beliefs include Jonah and the giant fish (no fish could swallow a human whole, much less keep him alive in its stomach for three days) and the tale of Noah's Ark, which defies not only geology but reason (how could all of Earth's species, including dinosaurs, stay alive for a year on an ark with a single window?).

But not all beliefs are negotiable. For Christians, the story of how sin came into the world through Adam and Eve, and was expiated by Christ's death, is vital. It is the fundamental belief of Christianity, and rests critically on the historical existence of Adam and Eve and their status as the genetic ancestors of all humanity. Without their existence, and their transgression in the Garden of Eden, there would be no inherited sinfulness of humans, and without such sin there was no need for Jesus to appear on Earth, undergoing Crucifixion and Resurrection to redress our sins.

For other believers, the creation story as portrayed in Genesis, while perhaps not literally true, must somehow affirm the uniqueness of humans among all species—something that doesn't comport with purely naturalistic evolution. The involvement of God in the appearance of humans is, for many believers, nonnegotiable. After all, Genesis specifies that "God created man in his own image, in the image of God created he him; male and female created he them." This explains why, among those Americans who *do* accept human evolution, more than half of them believe that the process was guided by God, with the "guidance" usually nudging evolution toward our own species.

Finally, a critical claim of Mormonism is that Native Americans—including Moroni, the supposed creator of the golden plates that became the Book of Mormon—descend from a group of people who migrated to North America from the Middle East around 600 BCE.

Genetics, evolutionary biology, and archaeology show that all these claims are dead wrong. But because they are among the "nonnegotiables," they must somehow be saved. That is a job for accommodationists. Let's examine the claims and see how their believers buttress them against the winds of scientific evidence. We'll find that their defense demonstrates the failure of accommodationism: despite attempts to torture the facts into compliance with religious dogma, this strategy fails miserably. I concentrate on these cases not only because they involve crucial religious beliefs, but also because they involve my own areas of study: evolution and genetics. Further, more than any other area of science, it is biology in general and evolution in particular that are seen as being in direct opposition to scripture. With the possible exception of cosmology, which we'll discuss in the next chapter, religion can live happily with the modern findings of chemistry, physics, and nonevolutionary areas of biology like physiology and development.

Adam and Eve

The central lesson of Christianity is that sin was brought into the world by the transgression of Adam and Eve, the Primal Couple, and expiated by the Crucifixion and Resurrection of Christ, whose acceptance as savior removes the taint of sin. You can hardly call yourself a Christian without accepting these claims.

The idea that sin arrived with Adam and Eve's transgression originated in the epistles of Paul, but was transformed into dogma by Augustine and Irenaeus several centuries later. None of these writers doubted the historical existence of the Primal Couple. What could be clearer than Paul's declaration, "For since by man came death by man came also the resurrection of the dead. For as in Adam all die even so in Christ shall all be made alive"? As we've learned, Augustine, often praised for seeing Genesis as an allegory, actually regarded Adam and Eve as historical figures. Finally, the *Catechism of the Catholic Church* affirms a historical First Couple: "The account of the fall in Genesis 3 uses figurative language, but affirms a primeval event, a deed that took place at the beginning of the history of man. Revelation gives us the certainty of faith that the whole of human history is marked by the original fault freely committed by our first parents."

There's not much wiggle room here. And Americans as a whole take this doctrine literally as well: in a 2010 poll, 60 percent of them agreed with the proposition "All people are descendants of one man and one woman— Adam and Eve."

But science has completely falsified the idea of a historical Adam and Eve, and on two grounds. First, our species wasn't poofed into being by a sudden act of creation. We know beyond reasonable doubt that we evolved from a common ancestor with modern chimps, an ancestor living around six million years ago. Modern human traits—which include our brain and genetically determined behaviors—evolved gradually. Further, there were many species of proto-humans (all called "hominins") that branched off and died before the ancestors of our own species remained as the last branch. As many as four or five species of humanlike primates may have lived at the same time! Some of these extinct groups, like the Neanderthals, had culture and big brains, and were "modern" humans in all but name. Theologians, then, are forced to square the sudden incursion of sin with the gradual evolution of humans from earlier primates.

More important, evolutionary geneticists now know that the human population could never have been as small as only two individuals—much less the eight who rode out the flood on Noah's Ark. Since sequencing of human genomes became possible on a large scale, we can back-calculate from the observed genetic diversity in our species to find out roughly when different forms of human genes diverged from one another, and how many forms of a given gene existed at a given time. Because each human has two copies of each gene, this gives us a minimum estimate of how many *humans* existed at a given time. We've also been able to use genes to trace the path of ancient human populations as they spread from Africa throughout the world.

The genetic evidence tells us several things. First, the genes in all modern humans diverged from one another a long time ago—long before the 6,000 to 10,000 years estimated from scripture. We can, for example, trace all the Y chromosomes of existing males back to a single man who lived between 120,000 and 340,000 years ago. This individual is often called "Y-chromosome Adam." But that's a bit misleading, for although all the Y chromosomes of modern humans descend from this one individual, the *rest* of our genome descends from a multitude of different ancestors who lived at various times ranging from 10,000 to about 4 million years ago. Our genome testifies to literally hundreds of "Adams and Eves" who lived at different times—a result of the fact that different parts of our DNA were inherited differently based on the vagaries of reproduction and the random division of genes when sperm and eggs are formed.

The observations that different parts of our genomes have different ages, some going back millions of years, and that they come from different ancestors, completely dispel the biblical date of human origins and the idea that all of our DNA was bequeathed by a Primal Couple.

But the evidence is even stronger, for we can also back-calculate from DNA sequences the *size* of human populations at different times in the past. And we know that when our ancestors left Africa between 100,000 and 60,000 years ago to colonize the world, the size of the migrating group dropped to *a minimum of 2,250 individuals*—and that's an underestimate. The population that remained in Africa stayed larger: a minimum of about 10,000 people. The total number of ancestors of modern humans, then, was not two but over 12,000 individuals. This is a very strong scientific refutation of the Adam and Eve scenario.

And it puts Christians in a tight spot. If there were no Adam and Eve, then whence the original sin? And if there was no original sin transmitted to Adam's descendants, then Jesus's Crucifixion and Resurrection expiated nothing: it was a solution without a problem. In other words, Jesus died for a metaphor.

The scientific data have vexed many Christian theologians. Conservatives like the Southern Baptist Albert Mohler are predictably outraged:

> The denial of an historical Adam and Eve as the first parents of all humanity and the solitary first human pair severs the link between Adam and Christ which is so crucial to the Gospel. If we do not know how the story of the Gospel begins, then we do not know what that story means. Make no mistake: a false start to the story produces a false grasp of the Gospel.

Mike Aus, a liberal Protestant pastor, eventually left the church when he realized that Christian doctrine on Adam and Eve didn't square with evolution:

> Really, without a doctrine of original sin there is not much left for the Christian program. If there is no original ancestor who transmitted hereditary sin to the whole species, then there is no Fall, no need for redemption, and Jesus' death as a sacrifice efficacious for the salvation of humanity is pointless. The whole *raison d'être* for the Christian plan of salvation disappears.

Nevertheless, the Catholic Church continues to affirm the historicity of Adam and Eve and their original sin, its position unlikely to be reversed soon. Pope Pius XII's encyclical from 1950, explicitly denying multiple ancestors of modern humans, remains church doctrine:

> For the faithful cannot embrace that opinion which maintains that either after Adam there existed on this earth true men who did not take their origin through natural generation from him as from the first parent of all, or that Adam represents a certain number of first parents. Now it is in no way apparent how such an opinion can be reconciled with that which

the sources of revealed truth and the documents of the Teaching Authority of the Church propose with regard to original sin, which proceeds from a sin actually committed by an individual Adam and which, through generation, is passed on to all and is in everyone as his own.

Of course, more liberal theologians have rushed into the breach with some solutions. But in the end they're worse than the problem, for the solutions are so clearly contrived that they can hardly be taken seriously.

I won't describe these in detail, but they take several forms. The first tries to save the notion of the two "ancestors" of humanity by suggesting that the ancestors were cultural rather than genetic. This is the "federal headship" (or *Homo divinus*) model floated by several theologians and religious scientists. The biochemist Denis Alexander, emeritus director of the Faraday Institute for Science and Religion at Cambridge University, explains:

> According to this model, God in his grace chose a couple of Neolithic farmers in the Near East, or maybe a community of farmers, to whom he chose to reveal himself in a special way, calling them into fellowship with himself—so that they might know Him as the one true personal God. From now on there would be a community who would know that they were called to a holy enterprise, called to be stewards of God's creation, called to know God personally. It is for this reason that this first couple, or community, have been termed *Homo divinus*, the divine humans, those who know the one true God, the Adam and Eve of the Genesis account.

You can guess the rest. One of the federal heads disobeyed God's commands, and that sin spread, like a virulent disease, to everyone else. Apparently the spread was not "vertical" (transmitted from parents to offspring) but "horizontal," like a virus passed between unrelated people. But this replaces one set of problems with another, for by trying to save some bits of scripture, it rejects others—with no good rationale. Catholic doctrine, for instance, maintains not only the existence of just two ancestors, but a vertical inheritance of sin, as if it were carried in the genes. Alexander's model is clearly motivated not by data but by the need to retain the credibility of a critical religious belief. In fact, to call it a "model" is an offense to science; it is instead an untestable, made-up story.

Another version, the "retelling model" of Alexander, gives up the attempt to base Adam and Eve on historical figures. Instead, it's said, the evolving human lineage became aware of God, but then for some reason unanimously rejected his presence and law. But this again leaves the origin of "sin" (whatever it may be) unexplained, and as Alexander himself notes, this model "evacuates the narrative of any Near Eastern context, detaching the account from its Jewish roots." Like the federal headship model, it takes Genesis as an allegory but the Gospels as literal truth.

The most "sophisticated" attempt to reconcile Adam and Eve with the data of genetics is that of the biblical scholar Peter Enns, an evangelical Christian. Accepting that the Bible was a historical document assembled by humans yet inspired by God, Enns simply sees Genesis as a metaphor for the creation of the Israelite nation. But he accepts a literal Crucifixion and Resurrection, as well as their redemptive effects.

As for Paul's absolute conviction that Adam and Eve existed, Enns hypothesizes that Paul was simply casting about for an Old Testament explanation for the decline of humanity that arrived with sin. As Enns argues, "One can believe that Paul is correct theologically and historically about the problem of sin and death and the solution that God provides in Christ without also needing to believe that his assumptions about human origins are accurate. The need for a savior does not require a historical Adam." While that may appeal to more liberal sentiments, it's still a made-up reconciliation that faces big theological problems. If "sin" is merely our evolved tendency to be greedy, aggressive, and xenophobic, then God, who either foresaw or directed evolution, becomes responsible for sin. That's unpalatable to theists who believe that sin results from our free choice.

And Paul—as well as many famous theologians—is seen as wrong in some of his beliefs but right about others. That raises the cherry-picking problem, which becomes obvious when Enns argues that "Paul's handling of Adam . . . is appropriating an ancient story to address pressing concerns of the moment. That has no bearing whatsoever on the truth of the gospel." But it surely does, for the "truth of the gospel"—presumably the divinity,

Crucifixion, and Resurrection of Jesus—is supported by precisely the same kind of evidence that once buttressed Adam and Eve. On what basis can we reject one story but accept the other? Like other attempts to save Christian doctrine by doing an end run around science, even the most "sophisticated" attempt seems like a desperation move fueled by confirmation bias.

Mormonism and the Origin of Native Americans

A linchpin of Mormon theology is that the ancestors of Native Americans were in fact Israelites of four tribes—Nephites, Lamanites, Mulekites, and Jaredites—who came to the New World from the Middle East about twenty-six hundred years ago. A thousand years later, the Nephites and Lamanites clashed, and the sole surviving Nephite, Moroni, helped write the Book of Mormon, burying it in upstate New York as a collection of golden plates. Later, taking the form of an angel, Moroni pointed it out to Joseph Smith in 1827. The Book of Mormon clearly states that North America was devoid of people when the Middle Eastern tribes arrived, for they "possess[ed] this land among themselves." Generations of Mormon teachings and prophecies affirm that North America was occupied solely by these immigrants and their descendants.

But as with the existence of Adam and Eve, both genetics and archaeology have shown that the Middle Eastern origin of Native Americans is a fiction. The data are clear: Native Americans, like all native peoples in the New World, descended from East Asians—Siberians—who migrated over the Bering Strait roughly fifteen thousand (not twenty-six hundred) years ago. The estimate comes from the dating of settlements, from other archaeological and linguistic studies (Native American languages, for instance, bear no trace of Hebrew), and, most important, from genetics, which shows a close affinity between the ancestries of East Asians and Native Americans.

Mormon theologians have tried the usual evasions to reconcile their scripture with science. They have, for instance, declared that the migrants from the Middle East landed in Central America. That, however, doesn't work, because Central Americans are also closely related to East Asians. Some apologists claim that the DNA from Middle Eastern ancestors was lost through interbreeding with Native Americans already living in North

America. But no such squatters are mentioned in the Book of Mormon. In the end, the church simply punts, stating that "DNA studies cannot be used decisively to either affirm or reject the historical authenticity of the Book of Mormon." To do otherwise would be to admit that the Book of Mormon was false, at least in an important claim. But of course had we found substantial Middle Eastern DNA in Native Americans, the church would use that as strong support for their dogma. This is the usual double standard in using evidence—accept it if it supports your preconceptions, reject it if it doesn't—that distinguishes science from religion.

Theistic Evolution

On twelve occasions since 1982, the Gallup organization has polled Americans on their beliefs about human evolution. Those surveyed are given three statements and asked which comes closest to their view. The first is this: "Human beings have developed over millions of years from less advanced forms of life, but God had no part in this process." This is the way scientists see evolution, as a purely naturalistic process, although we don't use terms like "advanced" because all living species have lineages of the same length, extending back to the first species. The second alternative is unadulterated biblical young-Earth creationism: "God created human beings pretty much in their present form at one time within the last 10,000 years or so." The third choice is "Human beings have developed over millions of years from less advanced forms of life, but God guided this process." (Historically, about 6 percent of those surveyed reject all three answers.)

The last alternative, evolution guided by God, is called "theistic evolution." Although roughly 40 percent of Americans choose the pure creationist answer, between two-thirds and three-quarters of those who *do* accept human evolution prefer the God-guided version to the purely naturalistic one. Such theistic evolution infuses religion into science.

Theistic evolution is not just the belief of most science-friendly Americans, but has also been accepted by the Catholic Church, often said to be supportive of "Darwinian" evolution. Yet Darwin never saw a role for God in his theory, and neither do modern scientists. Nevertheless, human evolution, says the church, involved a deliberate intervention by God, who inserted something unique among animals—a soul—somewhere in our lineage. Pope Pius XII made this explicit in his encyclical *Humani Generis*:

> The Teaching Authority of the Church does not forbid that, in conformity with the present state of human sciences and sacred theology, research and discussions, on the part of men experienced in both fields, take place with regard to the doctrine of evolution, in as far as it inquires into the origin of the human body as coming from pre-existent and living matter—for the Catholic faith obliges us to hold that souls are immediately created by God.

There is, of course, no empirical evidence for either a soul or its unique presence in humans. It's a superfluous religious add-on to a scientific theory.

Now, it's not so much evolution as a whole that bothers religious people, but *human* evolution. Nobody seems to care whether squirrels or ferns evolved via a purely unguided and naturalistic process. Christians and Jews, however, see *Homo sapiens* as "made" in God's image. And while the meaning of that statement has been debated for millennia, it clearly sets us apart from other creatures.

Naturalistic evolution hits these views right in the solar plexus. To an evolutionary biologist, our lineage is simply one twig among millions on the great bush of life. Granted, we have special features like culture and a big brain, but we also lack features that make other species "special" (we can't photosynthesize, fly on our own, or hibernate). And there's nothing about our big brains—the source of our rationality, intelligence, and culture—that isn't consistent with natural selection acting on social primates.

Notice that I said *natural* selection. Darwin undoubtedly used that term not only to distinguish it from the *artificial* selection practiced by animal and plant breeders, but also to emphasize that the process, which he saw as his most novel idea, was purely natural.

Besides dethroning us as nature's pinnacle, evolutionary biology discomfits us in other ways. It implies that because species change purely as a result of random mutations having different abilities to propagate, there's no need for a creator. And because life itself probably arose via a similar process of "chemical selection" among collections of molecules, there is probably no sharp distinction between the *origin* of life and the *evolution* of life. Not only that, but natural selection and extinction seem like cruel and wasteful ways to "create" a world, producing additional headaches for theologians forced to explain why a loving God would create in such a way.

The blows keep coming. Evolution disproves critical parts of both the Bible and the Quran—the creation stories—yet millions have been unable to abandon them. Finally, and perhaps most important, evolution means that human morality, rather than being imbued in us by God, somehow arose via natural processes: biological evolution involving natural selection on behavior, and cultural evolution involving our ability to calculate, foresee, and prefer the results of different behaviors.

It's no wonder, then, that theistic evolution is centered squarely on the evolution of our own species. When John Scopes was convicted in Dayton, Tennessee, in 1925 in the famous "Monkey Trial," it was not for teaching evolution, but for teaching *human* evolution, for only the latter violated Tennessee's Butler Act. In Muslim countries, and even Western ones, Islamic schools may teach evolution, but nearly always with the caveat that humans were created specially by Allah.

Human exceptionalism was advocated even by Alfred Russel Wallace, the codiscoverer with Darwin of evolution by natural selection. Observing that the brains of modern humans can do far more things than could possibly have been favored by natural selection (music, playing chess, doing complex mathematics), Wallace concluded that "in [a modern human's] large and well-developed brain he possesses an organ quite disproportionate to his actual requirements—an organ that seems prepared in advance, only to be fully utilized as he progresses in civilization." Because evolution can't bestow individuals with features that become useful only in the future, Wallace concluded that the human brain could never have evolved by natural selection. He concluded that "the brain of prehistoric and of savage man seems to me to prove the existence of some power, distinct from that which has guided the development of the lower animals through their ever-varying forms of being." Here Wallace grants a single trait in nature—the human brain—an exception from pure naturalism. Because he was not conventionally religious, this view probably came from his immersion in mysticism and spiritualism that began several years earlier.

Theistic evolution and human exceptionalism aren't just the views of garden-variety believers, but are espoused today by some science-friendly theologians. Here are two modern theologians—first John Haught and then Alvin Plantinga—arguing that naturalistic evolution is both unpalatable and un-Christian:

> Religions can put up with all kinds of particular scientific ideas so long as these ideas do not contradict the sense that the whole scheme of things is meaningful. Religions can survive the news that Earth is not the center of the universe, that humans are descended from simian ancestors and even that the universe is fifteen billion years old. What they cannot abide, however, is the conviction that the universe and life are pointless.

> What is *not* consistent with Christian belief, however, is the claim that evolution and Darwinism are *unguided*—where I'll take that to include *being unplanned and unintended*. What is not consistent with Christian belief is the claim that no personal agent, not even God, has guided, planned, intended, directed, orchestrated, or shaped this whole process. Yet precisely this claim is made by a large number of contemporary scientists and philosophers who write on this topic.

As an evolutionary biologist, I'd respond that all the evidence points to unguided evolution, and even if that's distressing, it's the best inference we have. After all, we have to accept lots of things we don't like, including our own mortality.

Why is theistic evolution a failure of accommodationism? Before I explain, we should realize that "theistic evolution" is a semantic umbrella covering diverse and sometimes conflicting views differing in the assumed amount and nature of God's intervention. The least intrusive of these is a "let it roll" brand of deism, one that sees evolution as simply operating according to the physical laws created by God. (Advocates of this view differ in whether the origin of life required God's intervention.) Once the process is under way, God no longer interferes.

A more teleological interpretation sees evolution as *inherently progressive*. This was expressed by the theologian and former physicist Ian Barbour:

"The world of molecules evidently has an inherent tendency to move toward emergent complexity, life, and consciousness." The process driving evolution in this direction is often unspecified, but is somehow directed by God.

Barbour's view, one that is widespread, implies that humans were a built-in feature of evolution, one envisioned by God when he set up the process. In other words, big-brained humans, or similar humanoid creatures, were a *planned* result of evolution, and were therefore inevitable. Others also see the evolution of humans as inevitable, but argue that divine intervention wasn't needed—that there simply existed an open ecological niche for a conscious and rational animal, one capable of apprehending and worshipping the divine. And in time, one of the endlessly ramifying products of evolution would fill this niche.

Still others see God as having to interfere *sporadically* in evolution, guiding it in various ways we'll discuss below. Divine interventions are deemed necessary to ensure both the initial appearance of life and the eventual appearance of humans, for such matters simply couldn't be left to naturalism. And this view shades insensibly into intelligent design (ID), the modern version of creationism that, while accepting a limited amount of evolution within species, insists that garden-variety Darwinian evolution simply can't explain some "irreducibly complex" features like the blood-clotting system of vertebrates or the complicated whiplike tails (flagella) that propel some bacteria.

While ID arguments have been refuted by scientists, versions of theistic evolution keep popping up like heads on the Lernaean Hydra, for believers are tenacious. Yet all of these versions come perilously close to ID creationism. One of them is the Catholic Church's insistence that God intervened at least once in the human lineage to insert a soul. This remains church dogma, expressed by Pope John Paul II in his famous 1996 message to the Pontifical Academy of Sciences:

> With man, we find ourselves facing a different ontological order—an ontological leap, we could say. But in posing such a great ontological discontinuity, are we not breaking up the physical continuity which seems to be the main line of research about evolution in the fields of physics and chemistry? . . . But the experience of metaphysical knowledge, of

self-consciousness and self-awareness, of moral conscience, of liberty, or of aesthetic and religious experience—these must be analyzed through philosophical reflection, while theology seeks to clarify the ultimate meaning of the Creator's designs.

It's hard to see this "ontological discontinuity"—the endowment of humans with a metaphysical soul—as anything other than creationism. Granted, it may have been a one-time intervention, but it still mixes science with religion, weakening the claim that Catholicism is compatible with evolution. With respect to evolution, the position of the Catholic Church differs from biblical creationism only in the amount of God's intervention.

Finally, some theistic evolutionists hold a "constant tweaking" model: God interferes frequently in evolution, tugging it, like an errant dog that won't take to its leash, in prescribed directions. These could involve preserving endangered species, creating new mutations, or tinkering with genes or environments. These interventions have two features: they are undetectable, rendering them immune to scientific investigation, and they are invariably used to give God a way to ensure the evolution of humans. Kenneth Miller suggested that this could occur if God simply fiddled with the movement of electrons:

> Fortunately, in scientific terms, if there is a God, He has left himself plenty of material to work with. To pick just one example, the indeterminate nature of quantum events would allow a clever and subtle God to influence events in ways that are profound, but scientifically undetectable to us. Those events could include the appearance of mutations, the activation of individual neurons in the brain, and even the survival of individual cells and organisms affected by the chance processes of radioactive decay.

It's ironic that Miller, who has produced some of the most compelling and convincing arguments against intelligent design, finally winds up touting God as using quantum mechanics to guide evolution. In this way he's camping on the outskirts of creationism. And why would God want to act in a "clever and subtle" (i.e., sneaky) way? Why is that better than creating humans de novo? The only advantage of Miller's theory is that God's interventions are conveniently undetectable.

Theistic evolution fails to harmonize science with religion because it pollutes evolution with creationism, positing interventions by God that are either scientifically refuted or untestable—and therefore superfluous. That is why, though eagerly embraced by the American public, theistic evolution has been completely rejected by scientists. Imagine if we had equivalents in other fields, like "theistic chemistry," proposing that God undetectably forges bonds between molecules, or "theistic gravity," claiming that the attraction between objects is maintained by a Ground of Being. Nobody, even believers, would take this seriously. The only reason why theistic *evolution* has gained traction is because it's politically expedient (scientists don't mind it because it gives religious people a foot in the evolution camp), and because for believers it removes some of the sting from naturalistic evolution. Believers don't propose the notions of theistic chemistry and physics only because those fields don't conflict with scripture. Only biology has theories that strike down the human exceptionalism touted in sacred texts.

But theistic evolution is also riddled with scientific problems. The big one is that despite adherents' claims that mutations in our DNA are biased in a given direction (i.e., are "nonrandom"), there is no evidence that useful mutations crop up more often when the organism "needs" them. Mutation would, for instance, be nonrandom if mammals moving to a colder environment experienced relatively more mutations producing longer fur. But there's no evidence for that. As far as we know, the mutational process appears to be "indifferent" (a term I prefer over "random"): errors occur in an organism's DNA regardless of whether they'd be good or bad for its survival and reproduction. While one could save theistic evolution by arguing that God-created mutations are undetectably rare—in effect, miracles—that's not a testable hypothesis. What we *have* shown, in experiments with microorganisms, is that no external force seems to be producing mutations in an adaptively useful way.

Further, evolution doesn't show the signs of teleological guidance or directionality proposed by theistic evolutionists. Evolutionary biologists long ago abandoned the notion that there is an inevitable evolutionary march toward greater complexity, a march culminating in humans. If one considers all species together, the *average* complexity of organisms has certainly increased over the 3.5 billion years of evolution, but that's just because life began as a simple replicating molecule, and the only way to go from there is to become more complex.

Contrary to popular wisdom, complexity isn't always favored by natural selection. If you are a parasite, for instance, natural selection may make you *less* complex, for you can live largely off the exertions of another species. Tapeworms evolved from free-living worms, and during their evolution have lost their digestive system, their nervous system, and much of their reproductive apparatus. Yet tapeworms are superbly adapted for a parasitic way of life: they simply pump out eggs and let their host do much of the metabolic work.

It doesn't always pay to be smarter, either. For some years I had a pet skunk, who was lovable but didn't seem very bright: in fact, sometimes he seemed to be unaware of anything but food. I mentioned this to my vet, who put me in my place with a sharp retort: "Stupid? Hell, he's perfectly adapted for being a skunk!" Intelligence comes with a cost: you need to produce and carry that extra brain matter, and crank up your metabolism to support it. When this cost exceeds the genetic payoff, the brain won't get larger. A smarter skunk might not be a fitter skunk. There are many cases in which organisms have lost features, becoming simpler because such loss was favored by natural selection. Organisms that invade caves often lose their eyes, for it's no advantage to have a useless organ that, besides being easily injured, can divert resources from other parts of the body that are useful. Remember that the currency of natural selection is *reproductive output,* and sometimes reproduction is enhanced by evolution's removal of features that aren't useful.

Finally, we know of no natural or supernatural process that drives evolution in certain directions; in fact, sometimes natural selection can drive species extinct, by adapting them to environments that are vanishing. I suspect that polar bears will go this route as global warming proceeds.

When thinking about evolutionary "direction," we should remember that there are just two important evolutionary mechanisms, natural selection and genetic drift. Drift is simply random changes in the proportions of genes caused by the vagaries of reproduction: it's the genetic equivalent of flipping coins. If different forms of genes (say, those producing blue versus brown eyes) made no difference to your number of offspring, the proportions of those genes in a population would simply fluctuate at random. This process is nondirectional by definition, and cannot produce adaptation.

The other mechanism, of course, is natural selection, which *is* nonrandom and does promote adaptation. Selection produces changes in traits that give an organism a reproductive advantage in its *current* environment. Although that process can occasionally be directional, as in an evolutionary "arms race" when predators and prey evolve higher efficiency in killing and avoiding each other respectively, the directionality is due not to God but to environments in which there's only a single way to improve. When the climate becomes colder—and major glacial cycles occur every hundred thousand years or so—organisms must adapt to low temperature or face extinction. When things warm up, the evolutionary direction is reversed. If theistic evolution is to be a truly coherent theory, its proponents must do more than raise it as a theoretical possibility: they must explain what mechanism makes evolution directional, guiding it toward humans, and show us how and where God intervened in that process.

An important claim of many theistic evolutionists, whether or not they invoke God's intervention, is that the evolutionary appearance of humans on Earth was inevitable. But that argument also dissolves under scrutiny.

Was the Evolution of Humans Inevitable?

If science can make a plausible case that the naturalistic evolution of humans, or of creatures with similar mental faculties, was inevitable, then theistic evolutionists get a big break. In that case we'd no longer need to invoke supernatural intervention to produce our species, for humans, or something like them, would always appear after sufficient evolutionary time. This would produce, through a purely material process, just what theists need: a complex and rational creature who apprehends and worships God. (Let's call such creatures "humanoids.") That leaves the naturalism in biology but still produces the outcome theists want. It's important, then, to see how far science supports the notion of human inevitability. Indeed, if we can't show that humanoid evolution was inevitable, then the reconciliation of evolution and Christianity collapses, for if we're really the special objects of God's creation, our appearance must have been guaranteed by either God or nature.

How can science address the question of whether naturalistic evolution would always produce a species like ours? One way is to assume that there was a preexisting but unfilled ecological niche for a humanoid creature, and that evolution would eventually work its way into filling that gap. But scientists aren't at all sure whether there are "empty niches" that precede the evolution of the organisms that fill them. After all, some organisms create their own niches through their evolved behavior, so the niches evolve along with the organism. The classic example is the beaver, which, by evolving the ability to chew down trees and assemble them into a dam, created its own lake habitat and food reservoir, complete with an enclosed house. That niche didn't exist before beavers, but was created by their ancestors, and has affected their subsequent evolution.

Given the quirky history of life, it's impossible to predict what new creatures will evolve. Who would have predicted, for instance, that two groups of birds, one in the New World and the other in Africa and Asia (the hummingbirds and sunbirds, respectively), would independently evolve the ability to hover like helicopters before flowers, drinking the nectar with long beaks and tongues? And even if we can identify things that *look* like empty niches, we don't know if organisms have the physiological equipment, or the right mutations, to evolve a way of life that seems available and adaptive. There are no examples of snakes that eat vegetation, for instance, yet there are many snakes and a lot of grass and leaves. Can we assert with confidence that if we wait long enough, the evolution of grass-eating snakes is inevitable?

Still, in many cases organisms must adapt to a relatively unchanging environment, and so we can sensibly speak of at least some *aspects* of a niche, or way of life, to which animals and plants must adapt. Mobile organisms that live in the sea, for example, must evolve ways to swim and to get oxygen while in the water. The strongest evidence for such preexisting niches is the phenomenon of "evolutionary convergence," often invoked to support human inevitability.

The idea is simple: species often adapt to similar environments by independently evolving similar features ("convergences"). Ichthyosaurs (ancient marine reptiles), porpoises, and fish all evolved independently in the water, and through natural selection all acquired strikingly similar streamlined shapes. Complex "camera eyes" evolved separately in both vertebrates and squid. Arctic animals such as polar bears, arctic hares, and snowy owls are either permanently white or turn white in the winter, hiding them from predators or prey. This camouflage, too, evolved independently in each lineage.

Perhaps the most astonishing example of convergence is the similarity between some species of marsupial mammals in Australia and unrelated placental mammals that live elsewhere. The marsupial flying phalanger looks and acts just like the flying squirrel of the New World. Marsupial moles, with their tiny eyes and big burrowing claws, are dead ringers for our placental moles. Until it went extinct in 1936, the remarkable thylacine, the pouched Tasmanian wolf, looked and hunted like the conventional placental wolf.

Convergence tells us something deep about evolution. There must be at least *some* preexisting "niches," or habitable environments, that call up similar evolutionary changes in unrelated species. That is, starting with different ancestors and fueled by different mutations, natural selection can nonetheless mold unrelated creatures in very similar ways—so long as those changes improve survival and reproduction. There were niches in the sea (probably involving lots of nutritious marine prey) for mammals and reptiles, so porpoises and ichthyosaurs became streamlined. Animals in the Arctic improve their survival if they are white in the winter. And there must have been niches for small omnivorous mammals that glide from tree to tree.

Convergence is one of the most impressive features of evolution, and it is common: there are hundreds of cases, thoroughly documented in paleontologist Simon Conway Morris's book *Life's Solution: Inevitable Humans in a Lonely Universe*. But the subtitle gives a key to its thesis: Conway Morris is a devout Christian who sees humanoids as something that evolution would inevitably produce:

Contrary to popular belief the science of evolution does not belittle us. As I argue, something like ourselves is an evolutionary inevitability, and our existence also reaffirms our one-ness with the rest of Creation.

This view is echoed by Kenneth Miller:

> But as life re-explored adaptive space, could we be certain that our niche would not be occupied? I would argue that we could be almost certain that it would be—that eventually evolution would produce an intelligent, self aware, reflective creature endowed with a nervous system large enough to solve the very same questions we have, and capable of discovering the very process that produced it, the process of evolution. . . . Everything we know about evolution suggests that it would, sooner or later, get to that niche.

But my own understanding of evolution suggests otherwise. I see the proper answer to the question "Is the evolution of humanoids inevitable?" as "We don't know, but it's doubtful." There are in fact good reasons to think that the evolution of humanoids was not only *not* inevitable, but a priori improbable. The reason is this: although convergences are common features of evolution, there are at least as many *failures* of convergence. These failures aren't impressive simply because they involve species that are missing. Consider Australia again. Although there are many convergences between placental mammals and Australian marsupials, there are also many types of mammals that evolved elsewhere that have *no* equivalents among marsupials. There is no marsupial counterpart to a bat (that is, a flying pouched mammal), or to giraffes and elephants (large pouched mammals with long necks or noses that can browse on the leaves of trees). Most tellingly, Australia evolved no counterpart to primates, or any creature with primatelike intelligence. In fact, Australia has many unfilled niches—and hence many unfulfilled convergences, including that prized "humanoid" niche. If high intelligence was such a predictable result of evolution, why did it not evolve in Australia? Why did it arise only once—in Africa?

That raises another question. We recognize convergences because unrelated species evolve similar traits. In other words, the convergent traits appear in two or more species. But sophisticated, self-aware intelligence is a single-ton: it evolved just once, in a human ancestor. (Octopuses and dolphins are also smart, but they do not have the stuff to reflect on their origins.) In contrast, eyes have evolved independently forty times, and white color in Arctic animals several times.

While the convergence argument can support the view that some evolutionary pathways are more probable than others, that argument rests on the existence of similar traits that evolve independently in more than one group. It cannot, then, be used to claim that a feature *that evolved only once* (i.e., our complex mentality) was inevitable. The elephant's trunk, an intricate and sophisticated adaptation—it has more than forty thousand muscles—is also an evolutionary singleton, as are feathers. Yet you don't hear scientists arguing that evolution would inevitably fill the "long-proboscis niche" or the "feathered-animal niche." Conway Morris, Miller, and others proclaim the inevitability of humanoids for one reason only: their religion demands it.

The most famous proponent of the noninevitability of evolution was Stephen Jay Gould. In his book *Wonderful Life: The Burgess Shale and the Nature of History,* Gould argued that the only real way to test whether the evolution of any species (like humans) was inevitable would be to start evolution over and over again, replaying the "tape of life" to see if humans always appeared. That, of course, is impossible, for we're stuck with only one realization of the evolutionary process.

But there are other ways to judge whether evolution is repeatable in this way. One way is to understand how the process works, an understanding that, combined with some knowledge of physics, suggests that the tape of life would play out differently each time, even if started under identical conditions.

Like many biologists, Gould argues that evolution is "a contingent process." The way natural selection molds a species depends on unpredictable changes in climate, on random physical events such as meteor strikes or volcanic eruptions, on the occurrence of rare and random mutations, and on which species happen to be lucky enough to survive a mass extinction. If, for example, a large meteor had not struck the Earth sixty-five million years ago, contributing to the extinction of the dinosaurs—and to the rise of the mam-

mals they previously dominated—all mammals might still be small noctur-
nal insectivores, munching on crickets in the twilight. And there would be
no humans. Based on this contingency, Gould concluded the evolution of
humans was "a wildly improbable evolutionary event" and "a cosmic accident."

But is evolution *really* "contingent"? That depends on what you mean by
the word. Evolution is certainly *unpredictable*, because we don't know exactly
how the environment will change or what mutations will occur. But "unpredictable" is not identical to "not predetermined." Most scientists are physical
determinists, accepting that the behavior of matter, at least on the macro level
(stuff that humans can perceive), is absolutely determined by the configura-
tion and laws of the universe. We can't always predict the weather, for instance, but that's only because we can't know everything about what affects
climate, including temperature and winds at every spot on Earth. It's possible
that if we had perfect knowledge of such things—which now include the human behavior that contributes to global warming—we could accurately predict the weather years in advance (even with our present sophisticated
instruments, we can barely predict it a day in advance). Likewise, it was
certainly determined well in advance that the asteroid that snuffed out the
dinosaurs would strike the Earth about sixty-five million years ago in the
vicinity of the Yucatán Peninsula.

The point is that even if evolution is "contingent," that doesn't mean "it's
not determined in advance," but only that "we don't know enough to predict
it." It's likely, then, that the course of evolution is determined by the laws of
physics. And that might imply that those laws would, given identical starting conditions, always yield identical products—including humans.

But there's still one hitch, and it's an important one. It involves quantum
mechanics, which tells us that on the microscopic level of particles like
electrons or cosmic rays (mostly fast-moving neutrons), things are *not*
determined, but are *fundamentally and unpredictably indeterminate*. If
you took a lump of radioactive uranium, for instance, and could observe
when each atom decayed, and then restarted the whole scenario with the
same lump, you'd find that during the rerun different atoms would decay,
and you'd never be able to predict which ones. (The *ensemble* of atoms,
however, does obey statistical laws, so that the "half-life" of a radioactive

element—the time needed for half the atoms to decay—is always the same.) Thus although a large *group* of atoms decays at a constant rate, it's impossible to predict which atoms will go first and which later. Such statistical regularities but individual indeterminacy are characteristic of quantum mechanical phenomena, including radioactive decay.

The question of whether humans were inevitable, then, boils down to the question of whether evolution is repeatable and deterministic, and that can be further reduced to the question of whether evolution is affected by the genuine indeterminacy of quantum mechanics. And it most likely is—in two important ways. The first involves whether the Earth would exist in the first place if the Big Bang were repeated under the same starting conditions. The answer is almost certainly no. Rerunning the history of the universe would probably result in a general similarity to what we have now (perhaps similar numbers of stars and galaxies, for instance), but it's very unlikely that the Earth and Sun would exist as the same objects they do now. If we can't even repeat the appearance of our solar system after a replay of the Big Bang, then all bets are off: there's no assurance that life as we know it would evolve. One might argue that life would still evolve on *some* planets in the universe, but there's no guarantee that that kind of life would be humanoid—the "image of God."

But there's another part of evolution that's also subject to the vagaries of quantum indeterminacy: mutations. Mutations are molecular changes in the DNA, many of them errors that occur when DNA replicates, which by changing the genetic code produce the new forms of genes that fuel evolution. And some factors that produce mutations, like X-rays, cosmic radiation, or even the simple errors in pairing of the DNA double helix, are probably affected by unpredictable quantum-level events.

What this means is that if life began all over again, even on our primitive Earth, the mutations that are evolution's raw material would be different. And if the raw material of evolution differed, so would its products: all the species alive today. All it would take is a few different mutations occurring early in the history of life, for instance, and everything that followed might have been very different from what actually evolved.

The upshot is that if mutations are fundamentally indeterminate, a replay of evolution would likely give us an array of species very different from those we see today. And we couldn't be sure at all that humans would be among them. The only way around this conclusion is to abandon naturalistic evolution and invoke a god supervising the process, making the right midcourse corrections to ensure that humans appeared.

Putting this together, if we replay the tape of either cosmic or biological evolution, we simply can't make a rational and logical argument that the appearance of humanoids was inevitable—and we can make a good argument that it was not. Any other answer involves either wishful thinking or unscientific claims grounded in theology, like God-directed mutations.

In the end, theistic evolution is not a useful compromise between science and religion. Insofar as it makes testable predictions, it has been falsified, and insofar as it makes claims that can't be tested, it can be ignored.

Theological Problems with Theistic Evolution

Does evolution pose further problems for theology? Yes, and big ones. There is no obvious explanation, for instance, why an omnipotent and loving God who directed evolution would lead it into so many dead ends. After all, over 99 percent of the species that ever lived went extinct without leaving descendants. The cruelty of natural selection, which involves endless wastage and pain, also demands explanation. Wouldn't a loving and all-powerful God simply have produced all existing species de novo, as described in Genesis?

As we saw in the last chapter, the usual response is to transform the unpalatable necessities of evolution into virtues, as does the Catholic theologian John Haught:

> The idea that secondary causes [natural selection], rather than direct divine intervention, can account for the evolution of life may even be said to enhance rather than diminish the doctrine of divine creativity. Isn't it a tribute to God that the world is not just passive putty in the Creator's hands but instead an inherently active and self-creating process, one that

can evolve and produce new life on its own? If God can make things that make themselves, isn't that better than a magician-deity who pulls all the strings, as theological "occasionalists" have supposed?

But one could easily make the opposite argument: that de novo creation is "better" because it avoids the suffering, waste, and extinction inherent in evolution. How does one weigh the value of creativity against the suffering of sentient creatures, including the several close relatives of modern humans, like Neanderthals, who went extinct?

Theistic evolution also comes in handy in theodicy, for you can use it, as does the evolutionist Francisco Ayala, to get God off the hook for all that "natural evil":

> The theory of evolution provided the solution to the remaining component of the problem of evil. As floods and drought were a necessary consequence of the fabric of the physical world, predators and parasites, dysfunctions and diseases were a consequence of the evolution of life. They were *not* a result of deficient or malevolent design: the features of organisms were not *designed* by the Creator.

The flavor of special pleading is strong here, for surely God, were he really omnipotent, could have designed a world whose physical fabric lacked floods, drought, and evolutionary suffering. And, of course, if God gets the credit for the adaptive mutations that led to humans, why is he exculpated for the maladaptive ones, like mutations that cause cancers, genetic diseases, and deformed children? If mutation were a process designed by God, there's no reason why the vast majority of mutations should be harmful— though that's exactly what you'd expect if the process were purely a naturalistic one involving random errors. Unsurprisingly, theologians like Alvin Plantinga have an answer to that one, too:

> But any world that contains atonement will contain sin and evil and consequent suffering and pain. Furthermore, if the remedy is to be proportionate to the sickness, such a world will contain a great deal of sin and a great deal of suffering and pain. Still further, it may very well contain sin

and suffering, not just on the part of human beings but perhaps also on the part of other creatures as well. Indeed, some of these other creatures might be vastly more powerful than human beings, and some of them—Satan and his minions, for example—may have been permitted to play a role in the evolution of life on earth, steering it in the direction of predation, waste and pain. (Some may snort with disdain at this suggestion; it is none the worse for that.)

It is astounding to see something like this coming from a respected philosopher. Not only do we encounter special pleading involving a God who makes animals atone for the sin of humans, but also the invocation of another source of evil: Satan. We are asked to believe, for instance, that the genetically based facial cancer wiping out Tasmanian devils could involve satanic manipulation of their chromosomes—innocent marsupials suffering horribly because of the sin of a primate. Snorting with disdain is in fact the proper response to this kind of ad hoc-ery, at least until Plantinga gives us some evidence for Satan.

It won't do for religious people to say that these answers are necessarily speculative because we have no idea how God works in evolution. If you admit that kind of ignorance, then you must also admit that we have no idea *whether God has anything to do with evolution*. It is curious that those who claim such firm knowledge about God's nature and works become silent when asked about God's methods.

The biggest problem with theistic evolution, as with all attempts to twist theology to fit new facts, is that it's simply a metaphysical add-on to a physical theory, a supplement demanded not by evidence but by the emotional needs of the faithful. It's what the philosopher Anthony Grayling calls an "arbitrary superfluity": the cosmological twiddling that Laplace rejected when he told Napoleon that the God hypothesis wasn't needed. And in fact, you find these superfluities only in evolutionary biology—and occasionally cosmology—for other sciences don't conflict with people's cherished beliefs. If you stretch science to include medicine, then we also have the arbitrary superfluity of seeing disease as a product of faulty thinking or spiritual error, a view that, as we'll see in the final chapter, has led many to reject science-based medicine and to suffer the consequences.

Finally, theistic evolution makes a common error of accommodationism: confusing *logical possibilities* with *probabilities*. Yes, it is logically possible that God started the evolutionary process, created the first organism, and then stood back to watch the action, or that he intervened from time to time, creating new organisms or mutations. But from what we know about evolution, that's unlikely. The process shows every sign of being naturalistic, material, unguided, and lacking divine assistance. To a scientist, theistic evolution fails because it requires that with one part of your brain—the "evolution" part—you accept only those things that are tested and supported by evidence and reason, while with the "theistic" part you rely on faith, assuming things that are either unnecessary or unevidenced. It's an unholy matrimony between science and religion, theology wearing a lab coat. We'll discuss the harms of this dysfunctional marriage in the last chapter, but some of the effects include the public misunderstanding of science (as in thinking that "theistic evolution" is scientific); the belief that religion can give us answers that currently elude science (e.g., why do the laws of physics permit the existence of life?); and the idea that there are "other ways of knowing," including revelation, that can yield truths about the cosmos. These are not just academic issues, for they have serious (and harmful) consequences for the real world: implications for morality, medicine, politics, ecology, and the general well-being of our species.

CHAPTER 4

Faith Strikes Back

When I was working as a pastor I would often gloss over the clash between the scientific world view and the perspective of religion. I would say that the insights of science were no threat to faith because science and religion are "different ways of knowing" and are not in conflict because they are trying to answer different questions. Science focuses on "how" the world came to be, and religion addresses the question of "why" we are here. I was dead wrong. There are not different ways of knowing. There is knowing and not knowing, and those are the only two options in this world.

—Mike Aus

The failure of accommodationism has led believers to engage with science in other than conciliatory ways. One way is for religion itself to don the mantle of science, claiming that there are some observations about nature that are best explained—or *only* explained—by the existence of a god. I call this strategy the "new natural theology," because it descends from earlier efforts to discern God's hand in nature. A related argument is that religion, like science, philosophy, and literature, is simply another "way of knowing" about the universe, possessing unique methods that yield valid truths. Some further argue that religion should get *credit* for science, because science was supposedly an outgrowth of faith—usually Christian faith.

When all else fails, believers find ways to denigrate science. Science is said to be an unreliable way to find knowledge (after all, scientific "truths" often change), is susceptible to misuse (read: atomic bombs and Nazi eugenics), and promotes "scientism," the view that science aims to engulf all other disciplines, forcing areas like history, literature, and art to become scientific or become irrelevant.

There are two more diagnostic signs that accommodationists have been pushed against the wall: their insistence that "science can't prove that God *doesn't* exist" (also known as "you can't prove a negative"), and their claim that science is just as fallible as religion because both ultimately rest on faith. As the psychologist Nicholas Humphrey has pointed out, denigrating science is often more appealing than making theological "god of the gaps" arguments, for "there must always be many extraordinary facts that could potentially discredit the conventional world-view [science], but relatively few facts that could provide positive support for a specific alternative [religion]. The project of doing down science was therefore always more likely to make headway than the project of bolstering a new kind of parascience."

These ideas, while not rare, are scattered through a huge literature, and some, like the "other ways of knowing" trope, are rarely discussed critically. In this chapter I'll examine the most common critiques of science. This is not just a tempest in an academic teapot, for by unfairly denigrating the field and trying to privilege indefensible "ways of knowing," the critiques do serious damage to science, and ultimately to society.

The New Natural Theology

No one infers a god from the simple, from the known, from what
is understood, but from the complex, from the unknown, and
incomprehensible. Our ignorance is God; what we know is science.
—Robert Green Ingersoll

While faith can be seen as belief without evidence, or belief with insufficient evidence to convince most rational people, that doesn't mean that the religious completely abjure evidence. If it supports their preconceptions, they'll accept it. Further, they'll even *seek* such evidence, and not just through revelation. The perpetual search for Jesus's tomb and Noah's Ark underline this yearning for evidence.

But there are also difficult problems that science hasn't yet explained—the origin of life and the biological basis of consciousness are two—and, given their difficulty, some may never be solved. These lacunae constitute openings for theology: opportunities to propose God as a solution. These are, of course, the famous "god of the gaps" arguments, and while the problem with proposing a god as a solution to obstinate scientific puzzles is obvious (science has a history of filling the gaps and displacing gods), supernatural solutions continue to appear whenever science faces a really hard problem. These arguments might in fact be seen as a backhanded form of accommodationism: the use of religion to supplement and complete the task of science. And they are at the same time a form of apologetics: if a god is the best explanation for a natural phenomenon, then that validates the existence of gods.

"Natural theology" represents the attempt to discern God's ways, or find evidence for his existence, by observing nature directly instead using revelation or scripture ("revealed theology"). It operates, or is supposed to operate, in a manner similar to that of science: aiming to show that for some of nature's puzzles, using God gives better answers than using naturalism. The philosopher Herman Philipse defines natural theology as "the attempt to argue for the truth of a specific religious view on the basis of premises that non-believers will be able to endorse, that is, without appealing to the alleged authority of a revelation." Natural theology is popular because, by offering God as the only reasonable solution to a scientific problem, it appeals to science-friendly believers as well as to theologians, both of whom realize that divine solutions can strengthen the faith of coreligionists and convert those on the fence.

Although natural theology has been practiced for centuries, it was especially popular in the West after science arose but before religion began to wane—between the seventeenth and the nineteenth centuries. And it was part of the scientific toolkit. As we've learned, Isaac Newton invoked the action of God to stabilize planetary orbits. In the absence of other natural explanations, that was convincing at the time.

Perhaps the most famous argument for natural theology was that of the clergyman/philosopher William Paley, whose 1802 book *Natural Theology* (subtitled *Evidences of the Existence and Attributes of the Deity*) argued that the "design" of animals and plants gave convincing evidence for a beneficent God. His most famous argument involved the "camera" eye of humans, an organ so complicated and composed of so many interrelated parts that it simply couldn't be explained by natural processes. As he said, "As far as the examination of the instrument goes, there is precisely the same proof that the eye was made for vision, as there is that the telescope was made for assisting it." Even the young Darwin initially found this argument convincing, though he later refuted it decisively by showing that natural selection alone could produce organs as complex as the eye.

It's a bit Whiggish to criticize early natural theology. At the time it could be seen as science, for it had the positive agenda of understanding nature using the best available explanations. Among the many phenomena thought to have a divine cause were mental illness, lightning, the origin of the universe, magnetism, and, of course, evolution. Natural theology was also scientific in that at least some versions went beyond post facto rationalizations to make predictions and testable claims. Creationism, for instance, makes claims about the age of the Earth, the absence of transitional forms between "kinds" of creatures, and so on. In 1859, it was a valid competitor to evolution, which is why Darwin spent so much time in *On the Origin of Species* showing how his new theories explained the data better than did creationism.

By the mid-nineteenth century, theology had finally lost its cachet as a form of science. This resulted largely from Darwin's own dismantling of the best argument for God ever derived from nature, as well as from the success of physics and medicine in replacing religious explanations with natural ones. Its decline was also hastened by philosophy, as Hume and Kant had given cogent arguments against miracles, deliberate design, and the logical arguments for God.

Yet natural theology has recently had a comeback. This is partly due to the rise of "intelligent design," which claims to identify evolutionary gaps fillable only by invoking an intelligent designer. While ID advocates argue that the designer is not necessarily the Judeo-Christian God—it could, they say, be an alien from another planet—this is disingenuous. The Christian roots of ID, and the private statements of its proponents, show that it's intended to replace the "disease" of naturalism with purely Christian metaphysics. ID is simply creationism gussied up to sound more scientific, in a vain attempt to circumvent U.S. court rulings prohibiting religious incursions into public schools.

Criticisms of "god of the gaps" arguments go back as far as ancient Greece, as we see from a famous statement by the physician Hippocrates of Cos:

> Men think epilepsy divine, merely because they do not understand it. But if they called everything divine they do not understand, why, there would be no end of divine things.

The theological dangers of this tactic were pointed out by Dietrich Bonhoeffer, writing from prison before the Nazis executed him for plotting against Hitler:

> How wrong it is to use God as a stop-gap for the incompleteness of our knowledge. If in fact the frontiers of knowledge are being pushed further and further back (and that is bound to be the case), then God is being pushed back with them, and is therefore continually in retreat. We are to find God in what we know, not in what we don't know.

Nearly everyone recognizes this problem, as did Ingersoll in the eloquent quote that heads this section. Yet believers continue to invoke God to explain unsolved scientific puzzles. This probably reflects a yearning not just for answers, but for answers that support the existence of one's faith. This might explain the equivocation of some religious scientists. For instance, Francis Collins, perhaps America's most visible scientist, is a strong opponent of intelligent design, and has repeatedly warned about using "god of the gaps" arguments:

> A word of caution is needed when inserting specific divine action by God in this or that or any other area where scientific understanding is currently lacking. From solar eclipses in olden times to the movement of the

planets in the Middle Ages, to the origins of life today, this "God of the gaps" approach has all too often done a disservice to religion (and by implication, to God, if that's possible). Faith that places God in the gaps of current understanding about the natural world may be headed for crisis if advances in science subsequently fill those gaps.

Indeed. Yet in the same book Collins violates his own prescription, arguing that God is the best explanation for not only the "fine-tuning" of the universe's physical constants, but also the innate moral feelings shared by most people.

The soothing feeling of having quasi-scientific evidence for your God is simply too alluring, leading even liberal believers to show the kind of ambivalence we saw in Collins. Here are some phenomena repeatedly cited by the new natural theology as evidence for God:

> The fine-tuning of the physical constants that allow our universe (and our species) to exist
> The existence of physical laws themselves
> The origin of life from inanimate matter
> The inevitability of human evolution
> Instinctive human morality (the "Moral Law" described above by Collins)
> The existence of consciousness
> The reliability of our senses at detecting truth
> The fact that the universe is even comprehensible by humans
> The amazing effectiveness of mathematics in describing the universe

When facing "scientific" arguments for God like these, ask yourself three questions. First, what's more likely: that these are puzzles only because we refuse to see God as an answer, or simply because science hasn't yet provided a naturalistic answer? In other words, is the religious explanation so compelling that we can tell scientists to stop working on the evolution and mechanics of consciousness, or on the origin of life, *because there can never be a naturalistic explanation*? Given the remarkable ability of science to solve problems once considered intractable, and the number of scientific phenomena that weren't even known a hundred years ago, it's probably more judicious to admit ignorance than to tout divinity.

Second, if invoking God seems more appealing than admitting scientific ignorance, ask yourself if religious explanations do anything more than rationalize our ignorance. That is, does the God hypothesis provide *independent and novel predictions* or clarify things once seen as puzzling—as truly scientific hypotheses do? Or are religious explanations simply stopgaps that lead nowhere? As I explained in *Why Evolution Is True*, Darwin's hypothesis for the change and diversity of life was accepted not just because it fit existing data but because it led to testable and verified predictions (e.g., where in the geological record would we find intermediates between reptiles and mammals?) and explained things that once baffled biologists (why do humans develop a coat of hair as six-month embryos, but then shed it before birth?). Intelligent design makes no such predictions or clarifications. Does invoking God to explain the fine-tuning of the universe explain anything *else* about the universe? If not, then that brand of natural theology isn't really science, but special pleading.

Finally, even if you attribute scientifically unexplained phenomena to God, ask yourself if the explanation gives evidence for *your* God—the God who undergirds your religion and your morality. If we do find evidence, for, say, a supernatural origin of morality, can it be ascribed to the Christian God, or to Allah, Brahma, or any one god among the thousands worshipped on Earth? I've never seen advocates of natural theology address this question.

We've already covered one item on the list above: the "inevitability" of humans having evolved, which on inspection doesn't seem so inevitable. As for the origin of life, we've made enormous progress in understanding how it might have happened beginning with inert matter, and I'd be willing to bet that within the next fifty years we'll be able to create life in the laboratory under conditions resembling those of the primitive Earth. That doesn't mean that it did happen that way, of course, for we'll probably never know. But reproducing such an event would falsify the religious claim that a natural origin of life is simply impossible without God. And the religious answer hasn't stopped the intense effort by chemists and biologists to find a naturalistic solution.

To the layperson, our consciousness seems scientifically inexplicable because it's hard to imagine how the sense of "I-ness"—and our subjective sensations of beauty, pleasure, or pain—could be produced by a mass of neurons in our head. Yet consciousness subsumes at least four phenomena: intelligence, self-awareness, the ability to access information (being unconscious versus "conscious"), and the first-person sense of subjectivity. Only the last—the so-called hard problem of consciousness—seems baffling, for it's difficult to imagine how a brain that can be studied objectively produces feelings that are subjective. Neuroscience has already made substantial inroads on the first three phenomena (brain interventions, both mechanical and chemical, as well as scans of brain activity, are putting together a neurological picture of what's required for these phenomena), and that field is already knocking on the door of the hard problem. Ultimately, its solution may elude us for one reason: we're using our limited cognitive abilities to tackle a research project that is hard even to frame. That doesn't mean that we should abandon this work, only that the most recalcitrant problem of science isn't the origin of life, or the origin of the laws of physics, but the evolution and mechanics of our brains and the minds they produce. This is the result of a peculiar recursion: we're forced to use an organ that evolved for other reasons to study how that organ makes us *feel*.

The "problems" of the laws of physics and the effectiveness of mathematics are connected, but only the first needs explanation. It's important to realize at the outset that the very term "laws of physics" is tendentious, for the word "law" implies a lawgiver, implying a creative god. But the laws of physics are simply *observed regularities* that hold in our universe, and I'll use "laws" in that sense alone.

And those laws are in fact a precondition of our existence, for without them we couldn't have evolved—or even existed as organisms. And by "we," I mean all species. If the laws of physics and chemistry varied unpredictably, our brains and bodies wouldn't work, for we couldn't have stable physiology, genetic inheritance, or the ability to reliably collect and process information about the environment. Imagine what would happen, for instance, if the impulses in our nerves, which depend on chemistry, traveled at widely variable speeds. Before we could pull our hand from the fire, or detect a predator, we might be burned or eaten. If we couldn't control the acidity of our blood within a fairly narrow range (one of the functions of our kidneys), we'd die. Nor could evolution have operated, for that process depends on environments being fairly constant from one generation to the next. If that constancy didn't exist, no species could last for long.

Thus, as with the "fine-tuning" problem of the laws of physics, which we'll discuss shortly, we can devise an "anthropic principle of biology": there *must* be stable laws of physics, or life wouldn't have evolved to the point where we could discuss these laws. Where natural theology comes in, as it does with physical laws, is the claim that the laws are "fine-tuned" by God to permit the existence of humans.

But if there are such laws, then the usefulness of mathematics is automatically explained. For mathematics is simply a way to handle, describe, and encapsulate regularities. As you might expect, there is in fact no law of physics—no regularity of nature—that has defied mathematical description and analysis. In fact, physicists regularly invent new types of mathematics to handle physical problems, as Newton did with calculus and Heisenberg with matrix mechanics. It's hard to conceive of *any* regularity that couldn't be handled by mathematics. So "the unreasonable effectiveness of mathematics in the natural sciences," as the physicist Eugene Wigner titled one of his scientific papers, simply reflects the regularities embodied in physical law. The effectiveness of math is evidence not for God, but for regularities in physical law.

One of the enduring goals of physics is to derive more fundamental laws from less fundamental ones: that is, to unite seemingly disparate phenomena under a single theory that reduces the number of subtheories. Classical Newtonian physics, for example, is now seen as a special case of quantum mechanics, and thermodynamics as a special case of statistical mechanics. The attempt to find a "theory of everything" unifying the four great forces of physics has been largely successful: so far only gravity has eluded union with the strong, weak, and electromagnetic forces. It is likely that this kind of amalgamation will continue on many levels, but it's also likely that in the end we'll reach a set of principles or descriptors that can be reduced no further. At that point we'll simply have to say, "These are, as far as we know, the irreducible laws of our universe." But what does it add to our understanding to say "and these laws are God's creation"? It adds nothing, but merely yields another unanswerable: where did that *God* come from?

But the existence of physical laws, even if we can't understand their constancy, raises a separate question. Even if those laws are *required* for our existence, why are they such as to *allow* our existence? To many believers, the answer is "Because God deliberately made them compatible with human life so that we, his most special creature, could evolve."

This brings us to the most pervasive—and, to the layperson, the most convincing—claim of natural theology: that of the "fine-tuning" of physical laws. After discussing it, I'll take up the remaining natural theological arguments for God: the "Moral Law" and our ability to hold beliefs that are true.

The Argument for God from "Fine-Tuning"

The "fine-tuning" argument for God, also known as the "anthropic principle," makes the following claim: Many of the constants of physics are such that if they differed even slightly, they would not permit the existence of our universe or the evolution and existence of humans who were supposedly the object of God's creation. Because we *do* exist, the values of these constants are too improbable to be explained by science, which suggests that they were adjusted by God to make life possible.

A full 69 percent of Americans support some version of this argument, agreeing that "God created the fundamental laws of physics and chemistry in just the right way so that life, particularly human life, would be possible." Even scientists are susceptible to the argument. Kenneth Miller has decried "god of the gaps" arguments, but relents a bit when those arguments involve physics:

> It almost seems, not to put too fine an edge on it, that the details of the physical universe have been chosen in such a way as to make life possible. . . . If we once thought we had been dealt nothing more than a typical cosmic hand, a selection of cards with arbitrary values, determined at random in the dust and chaos of the big bang, then we have some serious explaining to do.

The "serious explaining," of course, includes considering God.

But at the outset we must ask whether the constants of physics really do fall within a narrow range that permits human life. The answer is yes—but for only some of them. The constants that could not vary much from their measured values without making life as we know it impossible—and we'll need to discuss the meaning of "life as we know it"—include these: the masses of some fundamental particles (for example, protons and neutrons), the magnitude of physical forces (the strong and electromagnetic forces, as well as gravity), and the "fine structure" constant (important in forming carbon). If gravity were too strong, for example, planets could not exist long enough for life to evolve, nor would organisms be possible, as they would be flattened. The mass of the proton is 99.86 percent that of the neutron, and if it were even slightly smaller, stars like the Sun couldn't exist, for the nuclear fusion that powers their existence could not occur.

The observation that some constants must be close to what they are to make life physically possible is called the "weak anthropic principle." And it seems easy to refute this as evidence for God. If the constants didn't have those values, we wouldn't be here to measure them. *Of course* they must be consonant with human life.

But theists claim more than that, proposing what is called the "strong anthropic principle." This principle is simply the weak version with the added explanation that the constants fall within a narrow range of values because that's where God put them. As the argument goes, the values of those constants are *highly improbable* among the array of all possible physical constants, and because their values currently defy scientific explanation, that improbability is evidence for God. This is, of course, a "god of the gaps" argument, resting on our ignorance about what determines physical constants. But because this principle sounds authoritative and involves arcane issues of physics unfamiliar to the average person, it is often invoked by theists. It is in fact the most effective weapon in the arsenal of the new natural theology.

But do we really know that science will never be able to explain these constants of physics? No, of course not. We don't understand why they have the values they do, but we're making progress, and the conceptual gap that supposedly harbors God is closing.

Even the premise of "fine-tuning" is dubious, for we're not sure *how much* one can vary some of the constants without making life impossible.

Certainly the masses of protons and neutrons must be relatively close to their present value, but other physical constants need not be that precise. The cosmological constant and the entropy of the early universe, for instance, could have been substantially larger than they are without affecting our presence, for those changes would simply reduce the number of galaxies in the universe without affecting their fundamental properties. Stars could still exist, as well as planets that could harbor life. After all, God needed only one solar system to house the creatures who worship him, so why the superfluous billions of uninhabited planets?

Further, we don't know how improbable the values of the constants really are. Such a claim makes the crucial assumption that all values of constants are equally likely and can vary independently. It also assumes that there is no deep and unknown principle of physics that somehow constrains physical constants to have the values we see. Given our complete ignorance of the proportion of "physical-constant space" that could be compatible with life, there's simply nothing we can say about how improbable life is.

The fine-tuning argument for God gets even weaker because it includes the proviso "life as we know it"—that is, carbon-based life. But creatures sentient enough to perceive and worship God need not be the kind we know and love: our imaginations are limited by the life with which we're familiar. There could, for instance, be life based on silicon instead of carbon, life that could live at higher or lower gravity, or life that could have come into being instantly, rather than having evolved. After all, the Abrahamic God is omnipotent and so could fine-tune life for whatever universe he created, rather than fine-tune the universe for life.

Indeed, why should life be based on matter at all? To many people, God is a humanlike spirit, one with feelings and some sort of consciousness. Couldn't he create a race of similar immaterial but worshipful beings: bodiless minds that lack his powers? After all, any being with the power to determine physical constants could create any conceivable form of life, and God already is said to create souls, which are in effect bodiless minds. Why not just have a conclave of souls instead of material beings on Earth? The problem is that we simply don't know what forms of humanoid life—and here I mean "sentient, rational beings made in God's image"—are possible. Nor can theists explain why souls must have bodies.

Most important, there are other explanations for fine-tuning that don't invoke God. The simplest is that if we inhabit the only universe there is, we simply got lucky: that our universe had the right physical constants to permit and support life as we know it. In other words, in the bridge game of cosmology, we drew a nearly perfect hand—at least for carbon-based humanoid life.

The odds of this are immensely increased, however, if there is more than one universe: the equivalent of billions of hands of bridge being dealt at once. The concept of a "multiverse"—many universes that are independent of one another—falls naturally out of several current and popular theories of physics, including string theory and the idea of cosmic inflation: the very rapid expansion of the universe just after the Big Bang. If those theories are true, then there may well be multiple and independent universes. Further, the constants of physics will differ among those universes. Given that, it becomes probable that some universes will have the right physical constants to allow life as we know it, and lo, we happened to evolve in one of those. If you deal a huge number of bridge hands, one that's perfect, or close to it, becomes probable.

It's important to understand that the multiverse theory, the best current solution to the "fine-tuning" problem, is not, as some theists claim, a Hail Mary pass thrown by physicists desperate to avoid invoking God. Rather, it's a natural outcome of well-established theories. Now, it's not clear whether we can actually *show* that there are multiple universes, for they might be undetectable from our own. Still, physicists are beginning to devise ways to test their existence, and we've recently seen evidence for at least one of their preconditions: cosmic inflation. And even if multiverse theory is hard to test, the alternative "God theory" is *impossible* to test, for it makes no predictions.

Yet there are observations that to some physicists militate against God as the cosmic fine-tuner. One, as I mentioned above, is the needless largesse of the universe: the sheer number of places where life—at least "life as we know it"—doesn't or can't exist. Assuming that life can inhabit only planets and not stars, conservative estimates suggest about one hundred to two hundred billion planets in our own Milky Way galaxy, and about 10^{24} *planets*—a trillion trillion, or 1,000,000,000,000,000,000,000,000 of them—in our observable universe. Most of these are uninhabitable. If humans are the goal of God's creation, why the excess?

Further, there's an unnecessary excess not just of worlds, but also of particles. As far as we can see, among the twelve known types of elementary particles that make up matter (six quarks and six leptons), only four of them—the up and down quarks and the electron and its neutrino—are necessary for the existence of physical objects, including planets and humans. The remaining eight "matter particles" decayed away immediately after the Big Bang. They, as well as the mysterious "dark matter," don't have an obvious theistic explanation.

The excessive numbers of planets and particles are verified predictions of purely naturalistic theories of physics, but aren't obvious consequences of a God hypothesis. If "fine-tuning" is evidence for God, and natural theology a form of science, then it behooves theologians to explain why there are particles not needed for life. And while they're at it, can they tell us which physical constants are essential for life, and how narrowly they're tuned? Their inability to do so shows that the strong anthropic principle is simply a theological add-on, for invoking God adds nothing to our understanding.

Several other observations go against the theistic explanation for fine-tuning. The first is that human tenure on Earth will end when the Sun, expanding in its death throes, will vaporize the Earth in less than five billion years—assuming that we don't destroy our planet before that via nuclear war or global warming. Perhaps humanity can be saved by a mass migration to other planets, but that doesn't solve the problem, for the universe itself will also end, perhaps in the same time frame, through the "heat death" that will happen when increasing entropy forces our entire universe to a temperature of absolute zero. Theists, however, may not be bothered by this, because some predict the imminent arrival of the End Times.

There's further evidence against the God hypothesis: our universe is almost completely inhospitable to any kind of life we know. Place a human at random somewhere in the universe, or even on a planet, and the chance that she'll die within seconds is overwhelming. Extreme temperatures, radiation, and lack of oxygen are hardly signs of a universe fine-tuned for our existence. In fact, the vast bulk of the Earth is similarly unsuitable: the oceans (70 percent of the surface of our planet), deserts, ice caps, and so on.

Is that a world designed for humans? Rather than assuming that the world was created for humans, the more reasonable hypothesis is that humans evolved to adapt to the world they confronted. Add to that the number of predators, diseases, and parasites that faced our ancestors, and still afflict us today, and you can reasonably conclude that God hasn't given us an especially comfortable home. Indeed, it seems that evolution has enabled us to barely hang on in a world determined to kill us.

Theistic explanations must also take on board the severely delayed appearance of humans, for the universe required ten billion years of physical evolution before life could even begin on our planet. If God is omnipotent, and wanted humans, why didn't he just create the universe, humans, and the species we need instantly, à la Genesis? This is not a fanciful question, but one that religionists must deal with. After all, if you claim to understand God's personality and intentions, can you plead ignorance on crucial matters like this? Herman Philipse argues that the lack of a good answer undermines all of natural theology:

> What reasons can God have had for preferring the long evolutionary route of the history of the cosmos and of life, if he wanted to create the human species? Theists should not answer this question by the traditional bromide that God's intentions are inscrutable for us. Such a move would annihilate the predictive power of theism, and thereby destroy the prospects of natural theology. Instead, they should come up with convincing reasons for God to take the evolutionary detour, assuming that he probably wanted to create humans in the first place.

Of course, theists can respond, and have responded, that evolution is simply more creative, more "self-realizing," than creation from nothing. But that doesn't explain why it took evolution so long to get off the mark. Similarly, responding to the physicist Sean Carroll's observations that the universe has features *not* optimized for human life, the theologian William Lane Craig simply claimed that God might have some previously unsuspected artistic talents:

> But why should we think of God on the analogy of an engineer? Suppose God is more like the cosmic artist who wants to splash his canvas with

extravagance of design, who enjoys creating this fabulous cosmos, designed in fantastic detail for observers. In fact, how do we know that there isn't extraterrestrial life somewhere in the cosmos that needs these finely-tuned parameters in order to exist? Or perhaps God has over designed the universe to leave a revelation of himself in nature, just as it says in the book of Romans, so that someday physicists probing the universe would find the fingerprints of an intelligent designer, who is incredibly intelligent, incredibly precise, in making this universe.

Such elaborate storytelling is the last resort of the theologian, and it's quickly discredited by the lack of evidence—the "fingerprints of God." But such special pleading renders the entire program of natural theology unfalsifiable and therefore unworthy of serious consideration. Nor does it give any credence to the Abrahamic God beloved of Craig. Even if one was convinced that the universe was fine-tuned by a deity, why couldn't it have been Allah, Brahma, or Quetzalcoatl? As Christopher Hitchens used to say of such a problem, all the work is still ahead of them.

We still don't understand why the laws of physics are as they are, but scientists will never be satisfied with the answer "Because God wanted them that way." They may never find an answer, but the difference between the natural scientist and the natural theologian is that scientists aren't satisfied that they've gotten the truth when they simply make up a hypothesis that can't be tested. And until they find a testable answer, scientists simply say, "We don't know."

The Argument for God from Morality

In his 1871 book *The Descent of Man, and Selection in Relation to Sex,* where Darwin first applied his theory of evolution by natural selection to humans, he did not neglect morality. In chapter 3, he floats what can be considered the first suggestion that our morality may be an elaboration by our large brains of social instincts evolved in our ancestors:

The following proposition seems to me in a high degree probable— namely, that any animal whatever, endowed with well-marked social instincts, would inevitably acquire a moral sense or conscience, as soon as its intellectual powers had become as well developed, or nearly as well developed, as in man.

A century later, the biologist Edward O. Wilson angered many by asserting the complete hegemony of biology over ethics:

> Scientists and humanists should consider together the possibility that the time has come for ethics to be removed temporarily from the hands of the philosophers and biologicized.

Wilson's statement, in the pathbreaking book *Sociobiology: The New Synthesis,* really began the modern incursion of evolution into human behavior that has become the discipline of evolutionary psychology. In the last four decades psychologists, philosophers, and biologists have begun to dissect the cultural and evolutionary roots of morality.

This effort is just at the beginning, for it involves laborious studies of animal behavior, neuroscience, and psychological tests of both primates and humans, but it's already starting to pay off. And the naturalistic study of morality has, more than any other scientific endeavor, deeply disturbed theists. Many of them can accept the Big Bang, and even evolution, but if there's any part of human behavior that religion has arrogated to itself, it's morality. And so in the face of scientific progress, religion continues to maintain that both the source and the nature of human moral judgments must involve God. One of the gaps that they see unfillable by science, but explainable by God, is the "moral instinct."

When you see someone drowning, and you're able to swim, your impulse is to jump in the water to save them. When you hear about someone like Bernie Madoff cheating others out of their life savings, you get angry and feel that he should be punished. If your child and his playmates were attacked by a dog, you'd rush to save your own child first.

There are many instant and instinctive behaviors and feelings like this, all reflecting what you think is the right or wrong thing to do. These are the so-called moral intuitions, defined by the psychologist Jonathan Haidt as "the sudden appearance in consciousness of a moral judgment, including an affective valence (good-bad, like-dislike), without any conscious awareness of having gone through the steps of search, weighing evidence, or inferring a conclusion."

To many religious people, neither secular reason nor science seems able to explain these instinctive judgments. The geneticist Francis Collins has publicly proclaimed the impossibility of a secular solution, and thus sees innate morality as evidence for God:

> But humans are unique in ways that defy evolutionary explanation and point to our spiritual nature. This includes the existence of the Moral Law (the knowledge of right and wrong) and the search for God that characterizes all human cultures throughout history. . . . DNA sequence alone, even if accompanied by a vast trove of data on biological function, will never explain certain special human attributes, such as the Moral Law and the universal search for God.

Besides the instinctive nature of morality, some of its manifestations, especially altruism and self-sacrifice, are seen as evidence for God because they too supposedly defy explanation by natural selection, or any secular hypothesis based on culture. As Collins notes, "Selfless altruism . . . is quite frankly a scandal to reductionist reasoning. It cannot be accounted for by the drive of individual selfish genes to perpetuate themselves."

The author Damon Linker draws the same conclusion from the case of Thomas Vander Woude, an American whose twenty-year-old son fell into a sewage-filled septic tank. Vander Woude dived in, rescuing his son but drowned in the process. And how can self-interest explain soldiers who save their comrades by falling on grenades or firefighters who risk their lives daily? Linker argues, "There are specific human experiences that atheism in any form simply cannot explain or account for. One of those experiences is radical sacrifice—and the feelings it elicits in us." Linker sees human altruism as a gift from not just God but the *Christian* God, who performed his own act of self-sacrifice by sending Jesus to an earthly death.

Of course, atheism, which is merely the lack of belief in gods, isn't responsible for explaining altruism and ethics, a task that properly belongs to philosophy, science, and psychology. And those areas have offered plenty of nonreligious explanations for the "Moral Law" and altruism. The explanations involve evolution, reason, and education.

But before we get to those explanations, we must question the premises. Do humans really have innate moral sentiments, and are they relatively uniform across people and cultures? If they are, does the uniformity reflect genetics, learning, or a combination of these, or are all such explanations insufficient, pointing to God? (I take "innate" to mean "part of a person's nature" rather than "hardwired genetically.") And how can we explain not only the moral acts, but also why we *approve* of them? If we can answer these questions plausibly without invoking divine intervention, then the default "god of the gaps" explanation is neither necessary nor parsimonious.

Certainly the idea of morality itself, as well as our tendency to classify some behaviors as moral or immoral, seems nearly universal. In his book *The Blank Slate*, Steven Pinker compiled a list, based on the work of the anthropologist Donald Brown, of "human universals": behaviors, beliefs, rules, and other aspects of social living seen in every culture surveyed. These include "distinguishing right from wrong," "moral sentiments," and the notion of "taboos." Now, this shows only that having moral *sentiments* seems universal, rather than the *particular* sentiments that Collins and Linker see as innate. But Brown and Pinker's list includes some of these as well: empathy; distinctions between in-groups and out-groups; the favoring of kin; the proscribing of murder, rape, and other forms of violence; the favoring of reciprocity; and the idea of fairness.

This list is, however, missing some behaviors we feel are innate. Note, for instance, that it doesn't mention either altruism or the approval of altruism. But more recent work shows a near universality of other moral sentiments. A survey of sixty "traditional societies," like the Hopi, Dogon, and Tzeltal, found that of seven key moral values specified *in advance* (obligations to kin, loyalty to the group, reciprocity, bravery, respect, fairness, and recognition of property rights), most were held by nearly all the societies, and every one was found in six "cultural regions" also identified in advance.

There are also moral instincts that appear nearly universal but can be revealed only by posing "moral hypotheticals": situations that never really occur but evoke instant and apparently instinctive moral judgments. Perhaps the most famous is the "trolley problem," discussed extensively by Philippa

Foot and Judith Jarvis Thompson, in which people are asked whether they'd throw a switch to derail a runaway train onto a sidetrack, killing one person walking on that track but saving five on the main track. Most people see throwing the switch as a moral act. In contrast, they strongly disapprove of an apparently similar act: throwing a nearby fat person onto the track to stop the train (you're assumed to be too light to stop the train yourself). Yet both situations achieve the same end: sacrificing one life to save five. Surveys show that there are many situations like this in which people's judgments concur regardless of their nationality, gender, ethnic group, or religion. All of this adds up to the idea that *feelings* of morality are widespread and that some moral judgments do seem innate, if by "innate" you mean "felt and exercised automatically." But that says nothing about the cause of the innateness, much less that it's God. For there are two naturalistic explanations as well: evolution and learning.

Now, by "universal" I mean "*nearly* universal," for there are clearly individuals who lack empathy, who cheat, or who feel no sense of shame. And, of course, large parts of some societies have engaged in wholesale immorality, such as Nazi Germany and parts of the antebellum United States that practiced slavery. Westerners consider genital mutilation and honor killings immoral, while many in the Middle East have no problem with these but see as immoral Western traditions like educating women and allowing them to dress as they want. Further, much of the behavior that we see today as unquestionably immoral—the disenfranchisement of gays, women, and members of other ethnic groups, the use of child labor, the torture of both humans and animals for amusement—were once an accepted part of Western society. In *The Better Angels of Our Nature*, Steven Pinker makes a strong case that since the Middle Ages most societies have become much less brutal, due largely to changes in what's considered moral. So if morality is innate, it's certainly malleable. And that itself refutes the argument that human morality comes from God, unless the moral sentiments of the deity are equally malleable.

The rapid change in many aspects of morality, even in the last century, also suggests that much of its "innateness" comes not from evolution but from learning. That's because evolutionary change simply doesn't occur fast enough to explain societal changes like our realization that women are not an inferior moiety of humanity, or that we shouldn't torture prisoners. The explanation for these changes must reside in reason and learning: our realization that there is no rational basis for giving ourselves moral privilege over those who belong to other groups.

But some of our moral behaviors, if not sentiments, almost certainly evolved. Evidence for that comes from finding parallels between the behavior of our own species and that of our relatives. The primatologist Frans de Waal and his colleagues have described many such parallels between primates and humans. Chimps, for instance, have tried to rescue other chimps drowning in the moats around their enclosures, and capuchin monkeys seem to show notions of "fairness," throwing a fit when they're given a cucumber but observe a monkey in the next cage getting a more desirable grape. Indeed, even rats have been shown to have a rudimentary empathy, freeing other confined and distressed rats by unlocking their cages—without receiving a reward. Curiously, rats do not show helping behavior toward strains that they are unfamiliar with, including, if they were fostered by members of a different strain, members of their own strain. This kind of in-group versus out-group discrimination is seen in humans and other primates.

Still, these parallels do not show that we share the same genes for moral feelings with our relatives, genes that would presumably have been passed down from our common ancestor. In experimental tests, for instance, capuchin monkeys show more "prosocial" (i.e., helping) behaviors than do chimps, though we're much more closely related to chimps than to capuchins. Orangutans are more closely related to humans than are capuchins, yet humans and capuchins, but not orangutans, show negative responses to unequal treatment. (In fact, even dogs and crows are more averse to inequities than are orangutans!) When you overlay animal "morality" on the known family tree of mammals, you find that behaviors that look "moral" have evolved *independently* in several lineages. This makes sense, for even closely related species can have very different social systems (bonobos and chimpanzees are one example), and different social systems select for different behaviors. Orangutans, for instance, are far more solitary than chimpanzees, and so would experience little natural selection for behaviors promoting group harmony.

This evolutionary convergence of "premoral" behaviors in unrelated lineages makes it more likely that some moral behavior evolved independently in our ancestors, particularly because we spent most of our evolutionary history living in the kind of small bands that would select for that behavior. That same independent evolution also militates against the God hypothesis—unless you think that God also instilled morality in rats, monkeys, dogs, and crows.

There is another way to determine how much of human morality, if any, is in our genes. If we have a hardwired and inborn morality, infants brought up without moral training will develop it automatically. Such experiments are, of course, unethical, but we can approximate them by observing the behavior of small infants who have had almost no moral instruction. And those infants show a modicum of empathy, but only toward familiar people, especially parents. They also show some rudiments of justice and fairness, but again directed mostly toward those in their in-group. To everyone else infants are selfish. The work of the child psychologist Paul Bloom and others has shown that infants are spiteful and do not even tolerate *equality* with strangers. They will, for instance, choose to receive one cookie while a nearby infant gets none, rather than the alternative in which both infants get *two* cookies. In other words, infants sacrifice their own well-being just to flaunt their superiority in acquiring goods. Bloom concludes that infants have limited innate empathy but little compassion and no altruism, traits that must be inculcated by parents and peers:

> There is no support for the view that a transcendent moral kindness is part of our nature. Now, I don't doubt that many adults, in the here and now, are capable of *agape*. . . . When you bring together these observations about adults with the findings from babies and young children, the conclusion is clear: We have an enhanced morality but it is the product of culture, not biology. Indeed, there might be little difference in the moral life of a human baby and a chimpanzee; we are creatures of Charles Darwin, not C. S. Lewis.

It's no fluke that among the cultural universals listed by Pinker is socialization by elders.

So what about altruism? This is a tricky subject, because "altruism" has both a common and a strict biological definition. The common definition is simply "helping someone without immediately expecting a reward." That would include, for instance, giving money to charity, helping an old person cross the street, taking care of your children, or, in the most extreme cases, sacrificing your life for someone else: the save-a-drowning-child, volunteer-fireman, and falling-on-the-grenade scenarios. The question is whether any or all of these defy explanation by culture, biology, or both.

The answer is no. While we're not exactly sure about the mix of culture and biology that determines our "moral" feelings and actions, we at least have plausible nonreligious explanations for all forms of altruism, from the least onerous to the most sacrificial.

For one thing, although some acts *seem* purely altruistic, they actually redound to the status of the person who does them. We benefit from having a reputation for generosity and by being public (and honest) about it. There's a reason why most wealthy donors insist that the art galleries, museums, and other institutions they endow prominently bear their names.

Further, some altruism is part of a tit-for-tat strategy in which you *expect* to be repaid someday. From mutual grooming in primates to helping a friend move, your act may seem altruistic at the time, but those relationships wouldn't last long if you were always a taker but never a giver. In fact, one can show from evolutionary game theory that such "reciprocal altruism" can easily evolve, especially in small, stable groups in which individuals can recognize and remember one another. It's no coincidence that that is precisely the scenario under which the vast bulk of human evolution took place.

We have, then, both cultural and evolutionary explanations for these less extreme forms of sacrifice. Altruism might have evolved as a reciprocal phenomenon that was adaptive for individuals in small groups (explaining why babies show preference for people that they recognize), and it might also be a way of burnishing one's reputation, explaining why we usually don't hide our generosity under a bushel. In fact, many aspects of cooperation and altruism are precisely those we'd expect if their rudiments had evolved. Altruism toward others is reciprocated most often when many people know about it, but often isn't when you can get away with free riding. Humans have sensitive antennae for detecting violations of reciprocity, they choose to cooperate with more generous individuals, and they cooperate more when it enhances their reputation. These are signs not of a pure,

God-given altruism, but of a form of cooperation that would evolve in small bands of human ancestors. Three other "cultural universals" that support this idea are concern for one's self-image (and the hope that it's positive), feeling shame for having transgressed, and recognizing other individuals from their faces.

And around that evolutionary nucleus could accrete, via culture alone, altruism that truly gets *no* reciprocation: giving to charities or the homeless anonymously, administering CPR to a fallen stranger, or simply waving someone ahead in traffic. As Peter Singer explains in his book *The Expanding Circle*, improvements in communication and transportation among populations have hijacked our "be nice to acquaintances" module, for our circle of acquaintances has widened, and we've learned that people are pretty much the same everywhere. (We've also learned that reciprocity goes beyond our band, as we now exchange goods and services with people across the globe.) And you can hardly demand that others treat you morally unless you do the same for them, for there's no way you can convince those others that, with all else equal, you deserve to be treated better than they are. From evolutionary roots, then, can grow a tree of altruism fertilized by culture.

Explaining altruism toward relatives is not a problem: since the 1960s evolutionists have understood how that works. Helping relatives is often not "true" altruism, for you stand to gain something from your sacrifice: the propagation of your genes. Behaviors that promote helping relatives can be favored by natural selection simply because they preserve copies of the genes that are carried in those relatives. If your expected genetic benefit—discounted by the degree of relatedness to those you're saving—exceeds the genetic cost of your "altruism," the behavior will evolve. In other words, evolution would promote your sacrificing your life with certainty if you could save more than two of your children, each of whom shares half your genes. And if your chance of dying (or losing future reproduction) were less than certain, then you might risk your life to save even a single child.

Such "kin selection" explains why parents care more about their own children than other people's. It explains why we care more about our children and our siblings than about our aunts, uncles, and cousins (we share fewer genes with more distant relations). And it's a good explanation for why Thomas Vander Woude would try to save his child. He didn't *know* that he would die, but simply had the impulse to save his child—something that's certainly built into us by natural selection. (Damon Linker argues that because Vander Woude's child had Down syndrome, and probably wouldn't reproduce, such altruism has no genetic payoff, but that's a profound misunderstanding of how evolution works. The evolutionary cue for helping is *seeing your child in danger*, not a genetic calculus of whether that child would reproduce—something our ancestors would never have known.)

The hardest cases for evolution are the ones I call *true* biological altruism. These are instances in which you make huge sacrifices, up to certain death (in evolutionary terms, you lose all future offspring), to help those who are unrelated. Such sacrifices reduce the expected number of offspring you'll have. These acts include the soldier-on-the-grenade scenario, the jobs of volunteer firefighters, who don't even get paid to risk their lives for others, and the altruism of people like Lenny Skutnik, who dived into icy waters in 1982 to save a drowning woman after a plane crashed in the Potomac River. Such scenarios cannot be explained by evolution alone, for behaviors that could reduce your reproductive output, and are not passed on through relatives you saved, would be pruned from the population. How can we explain these purely sacrificial acts?

There are several ways. One is simply the "expanding circle": a feeling of innate empathy we develop at the plight of strangers. In many cases, such as those of the firefighters and Skutnik, the altruist doesn't know for sure he'll be killed, for otherwise there's no reason to attempt a rescue. In cases like the grenade scenario, everyone is going to be killed anyway, and the sacrificial act may simply have piggybacked on our evolved tendency to help those with whom we're intimately familiar. It's no accident that soldiers who put their lives on the line for their fellows often call them "brothers."

The nonadaptive hijacking of sentiments that have evolved for other reasons is not rare. Animals that have their own litters will often adopt members of another species. I've just seen a video of a mother farm cat suckling a brood of ducklings along with her own litter. It's gone viral because it's so adorable, but there's a biological lesson here: when maternal hormones kick in, you might foster an animal that's not even of your own species, much less your own brood. This happens because the "adoption" option simply isn't common in nature, and natural selection has operated to promote the suckling of infants that happen to be nearby—which are almost invariably your own.

Such hijacking also occurs in wild species. Cuckoo birds are "nest parasites" that lay their eggs in the nests of other species. The young cuckoos then proceed to kill the host's own offspring, and the unrelated foster parents continue to feed the young cuckoos until they fledge. This grisly tactic is clearly adaptive for cuckoos: by never having to feed their own young, they get permanent babysitters and can have dozens of offspring, all raised by others. But it's very maladaptive for the host birds, who gain no benefit from raising a member of another species, and indeed, lose all their own offspring. The host's maternal instincts have simply been hijacked by cuckoos, and haven't counterevolved to recognize their strange offspring. If this happened in humans—and it does, in the case of people who adopt unrelated children—it would be seen as a case of extreme biological altruism. Yet nobody has argued that the "altruism" of cuckoo hosts, or the phenomenon of cats suckling ducklings, is inexplicable by science and therefore constitutes evidence for God.

In the end, there are ample secular explanations for altruism. While some of our moral sentiments surely derive from evolution in our ancestors, they are refined and expanded through culture: learning and communication. The genetic evidence comes from comparative work on other species, as well as studies of human infants. The cultural evidence, on the other hand, comes from seeing how many moral sentiments are learned, how variable they are across societies (even though readily instilled into infants adopted cross-culturally), and how much they have changed in just the past few centuries. In many ways human morality resembles human language: we're born with the propensity to acquire both, but the specific moral views we adopt, like the specific language we learn to speak, depend on the culture in which we're raised.

So while morality may seem "innate," at least in adults, that is easily explained as a result of genetic endowment modified by cultural indoctrination. Given that this "Moral Law," as Francis Collins puts it, does not defy science and psychology, there is no need to invoke some divine tinkering with human behavior.

But the God hypothesis for morality and altruism has its own problems. It fails, for example, to specify exactly which moral judgments were instilled in people by God and which, if any, might rest on secular reason. It doesn't explain why slavery, torture, and disdain for women and strangers were considered proper behaviors not too long ago, but are now seen as immoral. For if anything is true, God-given morality should remain constant over time and space. In contrast, if morality reflects a malleable social veneer on an evolutionary base, it should change as society changes. And it has.

The Argument for God from True Beliefs and Rationality

A more sophisticated argument for natural theology involves our ability to hold beliefs that happen to be true, ranging from "I'd better stay away from that lion" to "Tomorrow morning the Sun will 'rise.'" Some theologians argue that this ability can be understood only as a gift of God, and this argument has gained some traction in natural theology. To an evolutionary biologist, however, the "argument from true beliefs" seems so clearly wrong that one wonders why it's so popular. Possibly one reason is that theologians, who make this argument most frequently, either don't understand or don't accept the ability of both culture and evolution to give us a propensity to detect the truth. But the argument has also been promulgated by a highly respected philosopher of religion, Alvin Plantinga, and seems impressive because it's framed in arcane language, formal logic, and probability theory. But one need not know math or much evolution to see its problems.

But what is the argument? In short, Plantinga claims that humans could never have true beliefs about *anything* without God's intervention. Natural selection, he says, promoted only our ancestors' ability to leave copies of our genes through differential survival and reproduction. It doesn't give a hoot about whether our *beliefs*, such as they are, are true. Considering our cognitive faculties, for instance, Plantinga says:

> What evolution underwrites is only (at most) that our *behavior* is reasonably adaptive to the circumstances in which our ancestors found themselves; hence it does not guarantee mostly true or verisimilitudinous beliefs. Our beliefs might be *mostly* true or verisimilitudinous . . . but

there is no particular reason to think they *would* be: natural selection is
not interested in truth, but in appropriate behavior.

The most important "truth" that Plantinga thinks humans perceive is, of
course, the existence of the Christian God and Jesus, and the salvation that
can be attained only by accepting these deities. But he also claims that without God we wouldn't be able to perceive *scientific* truths, including those
involving biology and physics:

> God created both us and our world in such a way that there is a certain fit
> or match between the world and our cognitive faculties. The medievals
> had a phrase for it: *adequatio intellectus ad rem* (the adequation of the
> intellect to reality). The basic idea, here, is simply that there is a match
> between our cognitive or intellectual faculties and reality, thought of as
> including whatever exists, a match that enables us to know something,
> indeed a great deal, about the world—and also about ourselves and God
> himself.

Elsewhere he writes, "This capacity for knowledge of God is part of our
original cognitive equipment, part of the fundamental epistemic establish-
ment with which we have been created."

Plantinga argues that the naturalistic process of evolution is incapable of
producing a brain that apprehends the truth of *evolution*, much less of any
other idea. He therefore sees a conundrum: "What I'll argue is that naturalism is incompatible with evolution, in the sense that one can't rationally
accept them both." Plantinga himself waffles about whether *he* accepts evolution, for he seems to have a fondness for intelligent design.

What we have here, then, is the claim that a critical part of human cognition can't be explained by naturalistic evolution. Plantinga argues that our
real truth detector is a *"sensus divinitatis"* ("sense of divinity") installed in
humans—and no other species—by God, as part of our creation in his image. (This idea derives from John Calvin, a figure much admired by Plantinga.) This is clearly a "god of the gaps" argument, one that Plantinga sees
as forging a harmony between science and religion:

There is superficial conflict but deep concord between science and theistic religion, but superficial concord and deep conflict between science and naturalism.

I won't belabor the obvious objection that even if we had such a divinely installed *sensus,* it's not evidence for Plantinga's Christian God as opposed to any other god. While he sees the Christian divinity as a "basic belief," something as obvious as believing that you ate breakfast this morning, there's no reason why other deities could also have given us a *sensus divinitatis*—in fact, the firm belief of Muslims in Allah argues that the whole notion of a Christian *sensus* is insupportable.

But let's leave this theological ground and ask ourselves two questions: Do humans really have consistently accurate beliefs about the world? And, whether they do or not, can the truthfulness of *some* beliefs be explained by evolution and neuroscience?

The answer to the first question is yes: we generally see the external world accurately, and often have a good take on our fellow humans as well. But, importantly, we're also prone to all kinds of false beliefs. Steven Pinker lists a few:

> Members of our species commonly believe, among other things, that objects are naturally at rest unless pushed, that a severed tetherball will fly off in a spiral trajectory, that a bright young activist is more likely to be a feminist bankteller than a bankteller, that they themselves are above average in every desirable trait, that they saw the Kennedy assassination on live television, that fortune and misfortune are caused by the intentions of bribable gods and spirits, and that powdered rhinoceros horn is an effective treatment for erectile dysfunction. The idea that our minds are designed for truth does not sit well with such facts.

And we're particularly prone to self-deception. Psychological studies confirm that many of us tend to think we're smarter, better-looking, and more popular than we really are (for many examples, see Robert Trivers's book *The Folly of Fools: The Logic of Deceit and Self-Deception in Human*

Life). Such consistent self-overrating may in fact have an evolutionary basis, for it helps us get our way when dealing with other people. Nobody is more convincing than a liar who believes his own lies, so an inflated self-presentation in our ancestors may have gained them leverage with others—and a reproductive advantage.

Now add to our distorted self-image the prevalence of delusions and errors like climate-change denialism, as well as belief in UFOs, alien abduction, astrology, ESP, and so on (I'd add for believers, "all religions with the possible exception of the 'right one'—yours"), and we see that our species is vulnerable to all manner of false beliefs. We're deceived by optical illusions too, like Ted Adelson's stunning "checker-shadow illusion," in which a light square falling in the shaded part of a shadowed checkerboard is mistakenly perceived as being much lighter than a square lying outside the shadow—even though both are exactly the same shade. That too is probably a by-product of natural selection, which is likely to have given our visual system a way to detect and compensate for the effect of shadows on the color and hue of objects.

The skeptic and science writer Michael Shermer has written extensively about why people believe untrue things. The many reasons include confirmation bias (a preference for believing what we find comforting), ignorance of how probability works, resistance to change, and a penchant for indoctrination. Many of these tendencies could have been useful in the environment of our ancestors, an environment in which we no longer live. After all, our ancestors didn't encounter checkerboards in partial shadow, and didn't know probability theory.

To sum up, our brains are fairly reliable but hardly perfect organs for detecting truths. And many of these imperfections, pervasive in people of all faiths, can be plausibly understood as products of natural selection. So much for the argument that God gave us a reliable truth-detecting apparatus. In fact, if Plantinga's own *sensus divinitatis* were working properly, he wouldn't have accepted the scientifically discredited notion of intelligent design!

Of course, Plantinga has an answer for why there are so many atheists, Hindus, Jews, Muslims, and pre-Christian believers, like the Aztecs and ancient Egyptians, who were somehow unable to form true belief in the Christian God. The answer is that in those individuals the *sensus divinitatis* is or was "broken," dismantled by the effects of sin. Curiously, Plantinga argues that your broken *sensus* need not stem from your own sin:

> Were it not for sin and its effects, God's presence and glory would be as obvious and uncontroversial to us all as the presence of other minds, physical objects and the past. Like any cognitive process, however, the *sensus divinitatis* can malfunction; as a result of sin, it has been damaged. . . . It is no part of the model to say that damage to the *sensus divinitatis* on the part of a person is due to sin on the part of the same person. Such damage is like other disease and handicaps: due ultimately to the ravages of sin, but not necessarily sin on the part of the person with the disease.

Here we have an untestable explanation for an insupportable thesis. But we needn't mire ourselves in such arguments. Rather, let's address the real situation: humans are good at detecting some truths and poor at detecting others. Some of our beliefs are rational and supported by evidence, while others are not. Can we plausibly explain this using naturalism alone, including evolution?

Indeed we can. The first thing to realize is that humans aren't *born* with explicit beliefs; we're born with a brain molded by natural selection to *form* beliefs when the brain gets input from the environment. Some of those beliefs involve learning from parents and peers, others from empirical observation. It's easy to imagine that in our millions of years of living in small bands of hunter-gatherers, we, like many vertebrates, evolved a tendency to absorb what our parents taught us (surely one reason why religion persists). Your chances of surviving are a lot better if you benefit from the experience of adults rather than having to learn everything for yourself. We might have learned, for instance, to avoid large mammals that look like cats but not those that look like antelopes: those who didn't learn this distinction didn't become our ancestors. Other things, like which of our peers we can trust and which we should avoid, could have been figured out on our own. Still other rational tendencies—not beliefs, really, but adaptive behaviors—could have been directly produced by natural selection. Preference for kin and wariness toward strangers have obvious adaptive consequences, and would assume the status of "beliefs" at a certain age ("I believe that we should be wary of people we don't know").

While our view of the world is filtered through our senses, evolution has, by and large, molded those senses to perceive the world accurately, for there's a severe penalty to be paid for seeing things wrongly. That holds not only for the external environment, but also for the character of others. Without accurate perceptions, we couldn't find food, avoid predators and other dangers, or form harmonious social groups. And following those perceptions is indeed the pursuit of "true beliefs": beliefs based on evidence. Natural selection doesn't mold true beliefs; it molds the sensory and neural apparatus that, in general, promotes the formation of true beliefs.

But of course we've seen that not all of our beliefs are true. That's because while natural selection has given us a pretty good truth-detecting apparatus, that apparatus can also be fooled, as in the checker-shadow illusion. And religion could be one of the false beliefs that piggybacks on evolution. The anthropologist Pascal Boyer, for instance, proposes that religion began with the inborn and adaptive tendency of our ancestors to attribute puzzling events to conscious agents. If you hear a rustle in the bushes, you're more likely to survive (or get food) if you believe it came from another animal than from a gust of wind. These beliefs about conscious agents in nature can easily be transferred to things like lightning and earthquakes. Because our ancestors lacked naturalistic explanations for such things, conjectures about supernatural humanlike beings or spirits might follow. Studies of child development suggest that religious beliefs are not inborn—that we have no "God genes." Rather, as Boyer and others suggest, our evolved cognitive apparatus gives us a *propensity* to accept religious propositions such as God, the afterlife, and the soul, and those specific beliefs are learned.

It's no surprise that children in southern India come to believe in multiple deities, while those in Alabama in the divinity of Jesus—observations that directly contradict Plantinga's view that our *sensus divinitatis* was vouchsafed by the Christian God. Rather, religion is likely to be what Stephen Jay Gould and Elisabeth Vrba called an "exaptation": a feature that, while sometimes useful, was not itself the object of natural selection, but piggybacked on features that evolved for other reasons.

All in all, there's no reason to see science as incapable of explaining why our beliefs about the world are often true. It also has plausible explanations for beliefs that are false, explanations far more credible than Plantinga's view that our failure to detect truth indicates a *sensus divinitatis* broken by sin.

A related argument for God is that the human brain has abilities far beyond anything that would be needed by our African ancestors. We can build skyscrapers, fly to the Moon, cook elaborate dishes, and make (and solve) Sudoku puzzles. Yet such abilities could not possibly have been useful during nearly all of the period when our brain evolved. How then do we explain them? To some theologians, the answer is God.

As we learned earlier, the first one to raise this problem was the biologist Alfred Russel Wallace. Although a tireless and selfless promoter of evolution by natural selection (he called his book on the topic *Darwinism*), Wallace could not see how selection could produce the multifarious abilities of the human brain. Here's his argument, tinged with the paternalism of the nineteenth-century Englishman:

> We see, then, that whether we compare the savage with the higher developments of man, or with the brutes around him, we are alike driven to the conclusion that in his large and well-developed brain he possesses an organ quite disproportionate to his actual requirements—an organ that seems prepared in advance, only to be fully utilized as he progresses in civilization. A brain slightly larger than that of the gorilla would, according to the evidence before us, fully have sufficed for the limited mental development of the savage; and we must therefore admit, that the large brain he actually possesses could never have been solely developed by any of those laws of evolution, whose essence is, that they lead to a degree of organization exactly proportionate to the wants of each species, never beyond those wants—that no preparation can be made for the future development of the race—that one part of the body can never increase in size or complexity, except in strict co-ordination to the pressing wants of the whole.

In short, the brain seems to defy the idea that natural selection can't prepare organisms for environments they've never encountered. As a result, Wallace concluded that evolution could explain everything but a single organ in a single species.

We'll see in a moment that science has long since disposed of Wallace's teleology, but the idea of the overengineered human brain is still with us. It is in fact touted by Plantinga, who argues that because evolution can't explain our ability to do complex mathematics and physics, those aptitudes also come from God:

> These abilities far surpass what is required for reproductive fitness now, and even further beyond what would have been required for reproductive fitness back there on the plains of Serengeti. That sort of ability and interest would have been of scant adaptive use in the Pleistocene. As a matter of fact, it would have been a positive hindrance, due to the nerdiness factor. What prehistoric female would be interested in a male who wanted to think about whether a set could be equal in cardinality to its power set, instead of where to look for game?

This passage, written 140 years after Wallace's, is eerily similar.

But in the time separating Wallace and Plantinga, we've come to understand a lot about evolution, and we know now that our overengineered brain is not a puzzle for science. It's certainly true that our brain wasn't molded by natural selection for future contingencies. That is indeed impossible under naturalism, and Plantinga is also correct that doing mathematics would not have improved the fitness of our preliterate ancestors. But once the human brain attained a certain state of complexity—and it has to be pretty complex to handle language and the skills of living in bands of hunter-gatherers—it already had the ability to perform novel tasks that had nothing to do with evolution. Likewise, a computer designed to do certain things can, when its hardware becomes sufficiently complex, do things never envisioned by its maker.

Lest you think that this answer is special pleading, realize that similar phenomena occur in many animals. Crows, for instance, can use reason to solve complicated puzzles designed by humans (of course, there must be a food reward at the end). Parrots can imitate human speech, and even sing opera arias, while lyrebirds can imitate chain saws and car alarms. These talents are by-products of skills acquired for living in nature. Species often solve novel problems never encountered in nature, like the famous blue tits of Britain, who learned in the 1920s to open milk bottles delivered to doorsteps and guzzle the cream from the top. Chimpanzees have learned to crack nuts with rocks, to "fish" for termites by dipping chewed grass stems into the nest entrances, and to make sponges out of masticated leaves to soak up drinking water.

None of these behaviors could have been direct objects of natural selection; all were side effects of *other* aspects of the brain and body that were presumably the result of natural selection. If the human brain was overdesigned, then so were the brains of other animals, including our closest living relatives. That is no problem for biology, but it is for theists who claim that the *human* brain was uniquely overdesigned by a god—probably to apprehend and worship that god.

Before we leave this topic, it's worth noting that science is actually the best way to *correct* our false beliefs: beliefs that severed tetherballs fly away in spirals or that the Sun literally rises and sets. The elaborate cross-checking and doubt that pervade science, and the complicated instruments we've devised to supplement our senses, are all tools designed to check which of our beliefs are true.

Is Science the Only "Way of Knowing"?

All knowledge that is not the genuine product of observation, or of the consequence of observation, is in fact utterly without foundation, and truly an illusion.

—Jean-Baptiste Lamarck

One of the most common complaints of accommodationists and critics of "scientism"—the supposed overreaching of scientists that we'll discuss shortly—is that science has no monopoly on finding truth. In *The Language of God: A Scientist Presents Evidence for Belief,* Francis Collins asserts that "science is not the only way of knowing." The next sentence gives his alternative: "The spiritual worldview provides another way of finding truth."

But these "other ways of knowing," as they're commonly called, include more than spirituality and religion. Additional candidates are the humanities, social science, art, music, literature, philosophy, and mathematics. The whole panoply of "other ways" is touted not just by advocates of the humanities defending their bailiwick, but also by theists who want to use their own "ways of knowing"—faith, dogma, revelation, scripture, and authority—to buttress their claims about the divine.

I will argue that insofar as some of these disciplines can indeed yield knowledge, they do so only to the degree that their methods involve what I'll describe as "science broadly construed": the same combination of doubt, reason, and empirical testing used by professional scientists. Economics, history, and social science, for instance, can certainly yield knowledge. But religion doesn't belong in these ranks, for its "ways of knowing" can't tell us anything with assurance.

To evaluate any of these claims, we'll first need to define "truth" and "knowledge," which I'll admit can be tricky, for these concepts are historically mired in philosophical controversy. For consistency, I'll again use the *Oxford English Dictionary*'s definitions, which correspond roughly to most people's vernacular use. "Truth" is "conformity with fact; agreement with reality; accuracy, correctness, verity (of statement or thought)." Because we're discussing facts about the universe, I'll use "fact" as Stephen Jay Gould defined "scientific facts": those "confirmed to such a degree that it would be perverse to withhold provisional assent." Note that these definitions imply the use of *independent confirmation*—a necessary ingredient for determining what's real—and *consensus,* that is, the ability of any reasonable person familiar with the method of study to agree on what it confirms. Mormons confirm the verities of their faith by revelation and authority, but everyone else, including members of other faiths, withholds their assent. That's simply because there are no widely accepted observations that confirm Mormon dogma. It therefore fails to qualify as truth, scientific or otherwise. Finally, "knowledge" is simply the public acceptance of facts; as the *Dictionary* puts it, "The apprehension of fact or truth with the mind; clear and certain perception of fact or truth; the state or condition of knowing fact or truth." What is true may exist without being recognized, but once it is it becomes knowledge. Similarly, knowledge isn't knowledge unless it is factual, so "private knowledge" that comes through revelation or intuition isn't really knowledge, for it's missing the crucial ingredients of verification and consensus.

According to these criteria, science certainly finds truths and yields knowledge, for it includes not only procedures for generating theories about the universe, but the testability and repeatability that brings—or erodes—consensus. The consensus need not be *absolute:* there are a very few scientists who reject the truth of evolution. And there are still people who believe that the Earth is flat. But the rejection of evolution almost invariably rests on religious grounds, and the rejection of a round Earth is based on a kind of fanaticism that's blind to all evidence. While I'd hesitate to call these people "perverse," I'd certainly call their behavior irrational.

As I've noted, the conceptual tools of science (though not the title of "scientist") are available to everyone. I see science as a method, not a profession. Science construed in this broad way embraces all acts, including those of plumbers and electricians, that involve making and testing hypotheses. Indeed, that's exactly what we do when fixing our cars or trying to find lost objects by retracing our steps rather than looking elsewhere or praying for the answer. Any discipline that studies the universe using the methods of "broad" science is capable in principle of finding truth and producing knowledge. If it doesn't, no knowledge is possible.

Valid "ways of knowing," then, certainly include history, archaeology, linguistics, psychology, sociology, and economics, all of which, to greater or lesser degrees, use the methods of science. Historians, for instance, verify that Julius Caesar existed not only from the evidence of his own writings, but from writings by others, including contemporaries, who give consistent accounts, as well as from coins and statues made during his time. Holocaust denial, based largely on wish-thinking, has been refuted both in the courts and by historians armed with empirical evidence: interviews with survivors, guards, locals, and camp officials (and the agreement between their accounts); photographs of gas chambers and of the "selection" process at concentration camps; remains of the camps themselves; official Nazi documents; and population studies showing a severe attrition of European Jews during World War II. The evidence for a planned extermination of Jews, Romanis ("Gypsies"), gays, and others is so strong that Holocaust denialists can be classified as "perverse" under Gould's definition.

The social sciences are a bit less "scientific," because until recently the culture of these areas was less influenced by hard science, and the analyses and conclusions are usually still far less rigorous than those of, say, chemistry or biology. Nevertheless, sociologists can make testable predictions using lab studies or observations. One verified prediction (we could cite Marx as the source) is that decreasing the equality of income among members of a population will make it more religious. Psychologists often do experiments that are in every sense scientific: controlled, replicated, and analyzed statistically. And although economics is called the "dismal science," it becomes less so when conducting experiments about human greed or generosity, as in "behavioral economics," the field that fuses psychology and economics. Because microeconomic theories are hard to test—societies aren't often replicated—this area is perhaps the least scientific of the social sciences. Nevertheless, microeconomics has produced knowledge, including the shapes of supply and demand curves, the diminishing marginal utility of goods as they accumulate (the more doughnuts you have, the less you want another), and the relatively small effect on unemployment rates of extending unemployment benefits.

What about mathematics and philosophy? They're a bit different. Although they're useful *tools* for both science and rational thinking, they don't by themselves yield knowledge about the universe. (I'm not one of those who see mathematical truths as existing somewhere out there in the universe, independent of human cognition.)

The physicist Sean Carroll argues that to be scientific, a claim must have two qualities. It must be possible to imagine a world in which it is *false*, and it must, at least in principle, be testable by experiment or observation in the natural world. This is not true of mathematics. The Pythagorean theorem *must* be true in all worlds having the proper geometry, and is not tested but *demonstrated*. It is for this reason that mathematicians speak of "proving" a theorem, while scientists speak not of proof but of the strength of evidence. Nevertheless, it would be churlish to argue that the Pythagorean theorem, the value of pi as the ratio of two measurements of a circle, or Fermat's Last Theorem do not constitute "knowledge." They are indeed knowledge (or "truth")—knowledge not about the universe, but about the logical consequences of a series of assumptions.

Philosophy can produce a similar kind of knowledge, an understanding of the consequences that follow logically from certain premises. Although

Richard Feynman reportedly dismissed the value of philosophy to science with an infamous remark, "Philosophy of science is about as useful to scientists as ornithology is to birds," he was wrong on two counts. Philosophy of science *is* useful to scientists, and ornithology is useful to birds (many birders are conservationists). Philosophy, for instance, provides a rigorous framework for thinking about issues like consciousness, evolution, and evolutionary psychology, for finding fallacies in pseudosciences like creationism, and for interpreting science for the layperson. One of the great values of philosophy is its ability to find important logical errors. A good example is Plato's "Euthyphro Argument," which shows that, contrary to the claim of theists, most people derive their morality not from God's dictates but from secular thinking. This too seems a kind of knowledge.

But is morality itself a way of knowing? That is, are there objective moral "truths" to be discovered? I think not, for ultimately morality must rest on preferences: something seems "right" or "wrong" because it is either instilled in us by evolution, or conforms or fails to conform to how we think people should behave for their own good and for the good of their society. Some moral preferences are often nearly universal ("It is immoral to kill an innocent person"), but, as in the moral dicta of various religions, they often diverge among cultures. And when that happens, one must then justify why one act is moral and others are not. Such justification is invariably subjective. Religious people claim to discern right and wrong from revelation or scripture, which in the case of the Old Testament clearly approves of practices—slavery and the execution of adulterers and those caught working on the Sabbath—that we regard today as patently immoral. The Old Testament God also approved of genocide, ordering the complete extirpation—men, women, and children—of, among others, the Hittites, the Amorites, the Canaanites, the Perizzites, the Hivites, the Jebusites, and the Amalekites. These God-approved acts are quietly ignored by most believers.

Secularists like myself are often *consequentialists,* claiming that what is "moral" is what promotes a situation that you prefer, like harmonious societies, the well-being and flourishing of other people, and so on. And those preferences can (and must) be *informed* by observation and study—science. If you believe, for instance, that torture is wrong because it's incapable of extracting useful evidence that can save lives, such a belief can in principle be tested. But even if it can, that doesn't settle the issue, for people differ in how they weigh the saving of lives against inflicting pain on possibly innocent individuals, or against the detrimental effect that sanctioning torture has on a society's self-image and credibility. On what single scale can you objectively weigh the pain of someone who's tortured, the possible saving of lives from that pain, and the brutalization of society that might accompany the legal use of torture? Is there an "objective" answer to whether a third-trimester abortion is immoral, particularly if someone who opposes it has religious reasons?

It doesn't trivialize morality to argue that it is based on evolution and secular reason. After all, *some* principles of behavior are absolutely required for humans to live together in harmony, whether those principles be installed by natural selection or learned from interacting with others. It's worth adding that some people do feel that moral truths can be derived from science. Sam Harris, for instance, argues that what is moral is what increases "well-being," and that well-being can be measured. Most philosophers, however, agree that "ought" can't be derived from "is." I take their side. And if there are no objective moral truths, then morality isn't a way of knowing, but simply a guide to rational behavior.

This brings us to the realm of subjective experience, particularly the arts. Are painting, movies, literature, and music ways of knowing? (Remember that an affirmative answer still doesn't put religion in the same category.) Curiously, despite believers and academics who answer with a firm yes, the claim is rarely justified, and there is almost no discussion of which knowledge, if any, can be conveyed by the arts.

Clearly, works of art can tell us something about the character of the artist, about what he or she perceived, and about the type of human interactions in the society depicted or experienced by the artist. Just as clearly, the arts can stimulate our emotions or, more didactically, impart lessons about life. Two of my favorite pieces of fiction, F. Scott's Fitzgerald's *The Great Gatsby* and James Joyce's "The Dead," for instance, depict the futility of aspiring to wealth, repute, and true connections with others. *The Last Picture Show*, my favorite American film, shows people in a small town who, despite their best efforts, can't truly bond with one another. And my favorite foreign film, Kurosawa's *Ikiru* (*To Live*), shows us how a mundane and futile life can be redeemed by one simple act of kindness.

Such works can move us, and can even change us, but do they convey truth or knowledge? In the unforgettable last scene of *Ikiru,* the bureaucrat Kanji Watanabe, dying of cancer, sits in a swing in the playground he has built, happily singing as heavy snow falls all around. After a meaningless life of shuffling papers, he's finally done something *real,* bringing joy to children he'll never see. This depiction of redemption always brings me to tears. And Kurosawa surely intended us to feel that way. But we can disagree about whether *we* would feel redeemed in the same way. Although I can put myself in Watanabe's shoes, I'm not so certain that, given my temperament, I'd feel that building a playground could compensate for a lifetime of tedium.

We're often told that art and literature connect us to others, affirming our common humanity. But is that a truth about the universe? People are similar in some ways (we are devastated at the death of our partner) but different in others (only a fraction of us are intensely ambitious); art that points out these commonalities and disparities simply reinforces conclusions we learn empirically—from experience. It may do so in an artistic way—a way that makes us feel deeply—but it's not new knowledge. In many cases fiction or cinema immerses us in novel situations, challenging us to imagine what it would feel like to be in someone else's position. But would we really feel like Watanabe if we were sitting in his swing at the end of our lives? How do we know? Stimulating the imagination is not the same thing as imparting knowledge.

There are some exceptions. Before the existence of photography, sculptures or paintings could tell us how things appeared. We learn, for example, what the Hapsburgs looked like with their genetically inflated lower lips. Tolstoy's wonderful novella *The Death of Ivan Ilyich* was supposedly used in medical schools to help doctors understand what it feels like to die, and doctors still recommend it to teach empathy:

> Over a century after publication, *The Death of Ivan Ilyich* remains poignant to medical educators. It reminds us that as knowledgeable as we might be, it is still difficult to put ourselves in the patient's shoes. It reminds us that the same forces that distanced Ivan Ilyich from his caretakers continue to separate patients and physicians. . . . The goal of medical education should be to preserve the capacity to imagine a patient's suffering; we don't need to

"teach" empathy as much as we need to preserve the innate empathy the student brings. The study of medicine, the focus on disease and organ systems, can rob one of the qualities that brought one to medicine. *The Death of Ivan Ilyich* is a touchstone, a means of reconnecting with the sense of calling, and a reminder of how potent being fully present with the ill can be, a timeless therapeutic tool.

But the point about empathy has been made many times before: it is a "reminder," not a new discovery. In the end, Tolstoy's story doesn't convey new knowledge about death, but fictionalizes (albeit beautifully) knowledge presumably based on empirical observation. Indeed, Tolstoy could have been completely wrong about the feelings attendant on dying, and surely many people don't share Ivan Ilyich's final emotions. The story moves us because, if we've experienced the death of others, it generally rings true. Nevertheless, for doctors or medical students who lack empathy for terminal patients, the story may be a tool that prompts them to learn how dying people feel, just as philosophy helped physicists think more deeply about quantum mechanics.

In an essay on the cognitive value of art, the philosopher Matthew Kieran argues that whatever truth inheres in painting and literature comes from observing the real world:

> Consider the kind of putative insights we gain from fictions. Goya's *Disasters of War* (1810–1820) may convey war's horrors or Austen's *Pride and Prejudice* the dangers of self-regard, but do we learn such things from the artworks concerned? The idea that war is horrific or that pride comes before a fall is commonplace and trivial. If we already believe the message of such works then we cannot be said to learn anything from them. If we do not, then how could we learn from make-believe worlds that are not tied to truth about the real world? . . . For any truth claim conveyed through art we should look to the relevant mode of inquiry to check if it is warranted. We cannot learn, for example, from Austen about character—that is a matter for psychology.

I have asked literature professors and critics to give me examples of truths actually *revealed for the first time* by literature, rather than affirmed by it, and haven't received a single convincing answer. I would expect the same equivocation for music, painting, and other art, save for their ability (as in photography and painting) to tell us what something looked like. Art can prompt us to find truth, but in the end that truth must be based on reason and observation.

Note, too, that different works of art convey different—and sometimes diametrically opposed—truths, resembling the "truths" of different faiths. While scriptures and innumerable religious novels affirm the idea of a loving and omnipotent God, Voltaire's *Candide* satirizes it. Picasso's *Guernica* and the paintings of Goya convey the horrors of war, while innumerable Romantic paintings affirm its glory. Every *Piss Christ* by Andres Serrano—a crucifix immersed in a glass of the artist's urine—is offset by a worshipful *Isenheim Altarpiece* by Grünewald, a depiction of Christ's Crucifixion that I consider one of the world's most moving paintings. Surely Leni Riefenstahl's adulatory movies about Hitler and Nazism fall within the realm of "art" (and propaganda), but the truths we draw from them now differ from what the artist intended. What "knowledge" we get from such works is, at best, the knowledge of what the artist was trying to say.

Finally, it is clear that people of different cultures or different backgrounds will respond to art in different ways, gleaning diverse (and probably disparate) "knowledge." Would an Inuit draw the same lessons from *Moby-Dick* as Americans do now (I'm not counting the accurate descriptions of whaling, which came from Melville's research). Do we find the same "knowledge" in *Beowulf* that the Anglo-Saxons did ten centuries ago? What "truths" inhere in literature depend on one's background and culture, making them very different from the truths of science.

I'm certainly not arguing that art is worthless. Far from it. I derive immense satisfaction from books and paintings. But I appreciate them for their emotional resonance, for the depiction of other points of view, and for sheer aesthetics. Despite all this, I argue that art cannot ascertain truth or knowledge of the universe, simply because it lacks the tools for such inquiry. Insofar as art conveys knowledge, that comes from empirical observation and not through the artist's revelations, which, like the revelations of religious believers, tell us about the artist herself rather than the realities beyond her mind. Perhaps it's best to see art not as a way of knowing, but as a way of *feeling*, of giving us access to sumptuous beauty, personal validation, and, as with Buddhist meditation, a sense of solidarity and unity with others and the universe as a whole. Art intensifies and expands our subjective experience, and it's none the worse for that. And by stimulating our emotions and curiosity, it can be a tool, prompting us to search for real, verified knowledge.

When drawn from everyday life, such subjective experience serves as the last resort for adherents to the other-ways-of-knowing argument. The classic version, which I often hear from believers, is this: "I know my wife loves me"—supposedly a claim of knowledge beyond the ken of science. Like religious "truths," this assertion is said to be based on faith. Of course, someday science may indeed study love by measuring neurological activity or one's titer of hormones, and by correlating these things with one's claimed emotions, but until that day there's another scientific method: observation of behavior. As one of the commenters on my Web site argued:

We know the ways in which humans express romantic interest and/or love, and when we want to know if someone likes or loves us, we do so by inferring from his/her behavior.

What are teenage girls doing when they huddle around, dissecting how a particular boy is acting with a girl who is interested in him? They endlessly discuss all the clues, all the evidence in his behavior that suggests he is romantically interested . . . or not. My wife spends lots of time talking to her single friends. "Did he call you back the next day? Did he tell you he was going on that trip? Did he ask you to come?" etc. . . . all an analysis of the available observations looking for *evidence* of romantic feelings.

Why is it we think a guy—"Bill"—has a screw loose who turns up at a TV station with flowers ready to propose marriage to a pretty news anchorwoman who has never met him?

What about if "Susan" just went up to a guy on the street she'd never met, assuming the guy loves her and would immediately marry her?

What is the difference in these scenarios that mark them as "nut-case, irrational actions" vs. normal, loving relationship scenarios? It's that Bill and Susan are operating under a total lack of evidence for their belief that these people love them. So in direct contradiction to the claims of the religious apologist, we recognize mere "faith based" inferences of love as irrational, false, and non-indicative of how normal people conclude someone loves them.

"I'm hungry," my friend tells me, and that too is seen as extrascientific knowledge. And indeed, any feeling that you have, any notion or revelation, can be seen as *subjective* truth or knowledge. What that means is that *it's true that you feel that way*. What that doesn't mean is that the epistemic content of your feeling is true. That requires independent verification by others. Often someone claiming hunger actually eats very little, giving rise to the bromide "Your eyes are bigger than your stomach."

Which brings us to religion. In a way, this discussion has been a digression, for until now it has skirted the real issue: the ability of religion to find truth. The reason believers argue for "other ways of knowing" is simply to show that science has no monopoly on finding truth, and therefore that religion might muscle in alongside archaeology and history. But as we've seen, insofar as archaeology, economics, sociology, and history produce knowledge, they do so by using the methods of science broadly construed: verifiable, tested, and generally agreed-upon results of empirical study.

But even construing science broadly, one can't stretch it far enough to encompass religion. For even the most elastic notion of science doesn't include the methods that supposedly allow religion to gain knowledge: unverifiable authority of ancient books, faith, subjective experience, and personal revelation. As William James argued, it is the subjective and revelatory aspect of religion that gives it the most purchase: the feeling of certainty that religious claims are true. But when one has a religious experience, what is "true" is only that *one has had that experience,* not that its contents convey anything about reality. To determine that, one needs a way to verify the contents of a revelation, and that means science. After all, while some Christians accept the existence of Jesus because they have mental conversations with him, Hindus have mental conversations with Shiva, and Muslims with Allah. All the revelations in all the world's scriptures have never told us that a molecule of benzene has six carbon atoms arranged in a ring, or that the Earth is 4.5 billion years old. It is this asymmetry of knowledge that, despite religion's truth claims, makes its adherents embrace the fallacious claim that religion and science occupy separate magisteria.

The Scientism Canard

The other-ways-of-knowing claim is often coupled with accusations of "scientism": a behavior in which science or its practitioners are said to overstep their boundaries. Scientism is seen as an intrusion of science where it doesn't belong, an unwarranted invasion of philosophy, the humanities, ethics, and even theology. These are examples of what Stephen Jay Gould, in his NOMA argument, called the boundary violations of science. How dare, the critics say, science tell us *anything* about morality or aesthetics?

When we examine the behaviors described as "scientism," they're diverse and often unrelated. The physicist Ian Hutchinson sees it as an attempt to apply scientific methods to disciplines in which they're useless, trying to answer the "big questions" supposedly reserved for theology:

It is not merely the misapplication of techniques such as quantification to questions where numbers have nothing to say; not merely the confusion of the material and social realms of human experience; not merely the claim of social researchers to be applying the aims and procedures of natural science to the human world. Scientism is all of these, but something profoundly more. It is the desperate hope, and wish, and ultimately the illusory belief that some standardized set of procedures called "science" can provide us with an unimpeachable source of moral authority, a suprahuman basis for answers to questions like "What is life, and when, and why?"

The philosopher Susan Haack, on the other hand, sees scientism as science refusing to recognize its own limits, along with the problems this causes:

What I meant by "scientism" was . . . a kind of over-enthusiastic and uncritically deferential attitude towards science, an inability to see or an unwillingness to acknowledge its fallibility, its limitations, and its potential dangers.

Finally, the physician and bioethics expert Leon Kass characterizes scientism as the attempt to replace religion—and everything else—with science, a strategy that, he claims, could rend the very fabric of Western society:

> But beneath the weighty ethical concerns raised by these new biotechnologies—a subject for a different lecture—lies a deeper philosophical challenge: one that threatens how we think about who and what we are. Scientific ideas and discoveries about living nature and man, perfectly welcome and harmless in themselves, are being enlisted to do battle against our traditional religious and moral teachings, and even our self-understanding as creatures with freedom and dignity. A quasireligious faith has sprung up among us—let me call it "soul-less scientism"—which believes that our new biology, eliminating all mystery, can give a complete account of human life, giving purely scientific explanations of human thought, love, creativity, moral judgment, and even why we believe in God. . . . Make no mistake. The stakes in this contest are high: at issue are the moral and spiritual health of our nation, the continued vitality of science, and our own self-understanding as human beings and as children of the West.

The diverse notions of scientism have only one thing in common: they're all pejorative. In fact, the entry on "scientism" in *The Oxford Companion to Philosophy* begins:

> "Scientism" is a term of abuse. Therefore, perhaps inevitably, there is no one simple characterization of the views of those who are thought to be identified as prone to it.

And ends like this:

> A successful accusation of scientism usually relies upon a restrictive conception of the sciences and an optimistic conception of the arts as hitherto

practiced. Nobody espouses scientism; it is just detected in the writings of others.

But dire warnings like those of Kass are exaggerated. The dangers of scientism, no matter how you define it, are virtually nonexistent. To examine these supposed dangers, let's group the definitions into a few discrete categories. "Scientism" usually denotes one or more of the following four claims. First, science is the sole source of reliable facts about the universe; that is, it is the only reliable "way of knowing." Alternatively, scientism could mean that the humanities should be *subsumed* under the rubric of science. That is, areas like history, archaeology, politics, morality, art, and music should be viewed *only* through a scientific lens, and when possible should adopt the methods of science. Scientism could also refer to the idea that questions that can't be answered by science aren't worth considering or discussing. Such questions include those involving morality, ways to live, beauty, emotions, and, of course, religion. The most damning definition of scientism is the idea that scientists are arrogant, lack humility, and are reluctant to admit that their findings might be wrong.

As for the first claim, I've argued that science, construed broadly as a commitment to the use of rationality, empirical observation, testability, and falsifiability, is indeed the only way to gain *objective* knowledge (as opposed to subjective knowledge) about the universe. I've also argued that disciplines not normally considered "science" (like economics and sociology) can also produce knowledge when they use the methods of science. Finally, mathematics and philosophy produce a more restricted kind of knowledge: the logical results of assuming a set of axioms or principles. In the first sense of the term, then, most of my colleagues and I are indeed guilty of scientism. But in that sense scientism is a virtue—the virtue of holding convictions with a tenacity proportional to the evidence supporting them.

A few academics like Edward O. Wilson and Alex Rosenberg have indeed argued that eventually all areas of inquiry, including the humanities, will be not only united with science, but subsumed by it. The philosopher Julian Baggini argues for the futility of this takeover: "History, for example, may ultimately depend on nothing more than the movement of atoms, but you cannot understand the battle of Hastings by examining interactions of fermions and bosons."

This accusation is unfair. I've never heard a scientist claim that a knowledge of particle physics could give us insight into history. (Of course, many of us feel that if we had the unattainable perfect knowledge of every particle in the universe, we could in principle explain such macroscopic events.) A far more common claim is that many areas of the humanities, including politics, sociology, and literary scholarship, could be *improved* by insights from evolutionary biology and neuroscience. And really, who could disagree? Is there no room for empirical investigation in any of these areas—no way, for instance, that we could gain insights into human psychology by seeing it as partly a product of natural selection?

Indeed, archaeology, history, and sociology—even biblical scholarship—are increasingly informed by modern science. In a vigorous defense of that trend, Steven Pinker describes many other areas that have benefited from more rigorous, science-oriented approaches: evolutionary psychology is now a valid branch of psychology, articles in linguistics journals rely more on rigorous methodological inquiry, and data science promises to extract new information from economics, politics, and history. Naturally, some of the applications of science to these fields will be poorly motivated or executed, but that's not a problem of science itself, only of its misapplication. Presumably humanities scholars, like scientists, can recognize bad experimental design, flawed data analysis, or unsupported conclusions. And I'm certain that nearly all scientists agree with Pinker that our hope to help our colleagues in the humanities "is not an imperialistic drive to occupy the humanities; the promise of science is to enrich and diversify the intellectual tools of humanistic scholarship, not to obliterate them."

As for the claim that only scientific questions are worth discussing, I've met hundreds of scientists in my career, and I've never heard one say anything like that. Like all people, scientists can be arrogant and overbearing about their work, but so can novelists, artists, and historians! Nevertheless, more questions than we think can be *informed* by science, including those involving history, politics, the source of artworks, and issues of morality. After all, if you support the death penalty because you think it's a deterrent, or that certain offenders can never be rehabilitated, those opinions can be supported—or derailed—by empirical observation.

As we learn more about ourselves from evolution, psychology, and the neurosciences, more and more of the humanities become open to scientific study. Ian Hutchinson misses an important point when he judges beauty and emotionality as off-limits to science (accusations of scientism, of course, often come from the faithful):

> Consider the beauty of a sunset, the justice of a verdict, the compassion of a nurse, the drama of a play, the depth of a poem, the terror of a war, the excitement of a symphony, the significance of a history, the love of a woman. Which of these can be reduced to the clarity of a scientific description? . . . This is not a problem for science. It simply means that science is not able to deal with topics like these.

Not so fast. I'm confident that, someday, studies of neurology, genetics, and cognition will help us understand why some works of art move us and others don't, why some people are compassionate and others not, and why we see sunsets and waterfalls as beautiful but are repelled by wastelands. It's common to hear that love is a matter of "chemistry," but that's not just a metaphor, for surely the intense emotions that accompany love—sometimes verging on psychosis—are amenable to scientific analysis. Someday, for instance, we may be able to gauge the intensity (or even the presence) of love using neurology and biochemistry. That day may be decades away, but I'm not only sure that it will come, but just as sure that it won't stop poets and composers from writing paeans to love.

Like moral questions, there are many issues worth discussing that ultimately come down to matters of preference. How should I balance work versus play? Who was a better painter, Turner or Van Gogh? To which journal should I send my latest paper? I discuss things like these all the time with my fellow scientists. The notion that we disdain such questions is nonsense; even though we know there are no objective answers, we still might learn something.

Scientism is in fact a mug's game, a grab bag of disparate accusations that are mostly inaccurate or overblown. Nearly all articles criticizing scientism not only fail to convince us that it's dangerous, but don't even give any good examples of it. In the end, as Daniel Dennett argues, scientism "is a completely undefined term. It just means science that you don't like." Why don't people like it? Some in the humanities fear (without justification, I think) that science will render their disciplines passé, while religious believers labor under the misapprehension that tearing down science will somehow elevate religion.

Given its diverse meanings and lack of specificity, the word "scientism" should be dropped. But if it's to be kept, I suggest we level the playing field by introducing the term *religionism,* which I'll define as "the tendency of religion to overstep its boundaries by making unwarranted statements about the universe, or by demanding unearned authority." Religionism would include clerics claiming to be moral authorities, arguments that scientific phenomena give evidence for God, and unsupported statements about the nature of a god and how he interacts with the world. And here we find no lack of examples, including believers who blame natural disasters on homosexuality, tell us that God doesn't want us to use condoms, argue that the acceptance of evolution by scientists is a conspiracy, and insist that human morality and the universe's "fine-tuning" are evidence for God.

It would take volumes to answer all the criticisms leveled at science by believers and accommodationists. Here I'll briefly consider a half dozen of the most common claims.

Science Can't Prove That God Doesn't Exist

When an atheist debates a believer, the conversation often ends with the believer huffily asserting, "Well, anyway, you can't prove a negative." What he means is this: "No matter what arguments you raise against God, science can't demonstrate to me—or anyone—that he doesn't exist. For, as we all know, science can't prove that *anything* doesn't exist." That's a philosophical claim, one I hear quite often. Surprisingly, one claimant is the author and atheist Susan Jacoby:

> Of course an atheist can't prove there isn't a God, because you cannot prove a negative. The atheist basically says that based on everything I see

around me, I don't think so. Every rational thing I see and have learned about the world around me says there isn't a God, but as far as proving there isn't a God, no one can do that. Both the atheist and the agnostic say that.

Believers like the biologist Kenneth Miller, a Catholic, say the same thing:

> The issue of God is an issue on which reasonable people may differ, but I certainly think it's an over-statement of our scientific knowledge and understanding to argue that science in general, or evolutionary biology in particular, proves in any way that there is no God.

An alternative form of this argument is to claim that "the absence of evidence [for God] isn't evidence of [God's] absence."

Well, of course, if by "proof" you mean "absolute, unchangeable proof" (or in this case "absolute disproof"), Jacoby and Miller are right. Our understanding of reality—science's "truth"—is always provisional, and we can never rule out some kind of deity with absolute certainty.

But you can "disprove" God's existence in another way, by making two assumptions. First, the god under scrutiny must be theistic—one who has certain specified traits and interacts with the world. If you posit a deistic god who doesn't do anything, or a nebulous "Ground of Being" god lacking defined traits, then, of course, there's no way to get evidence either for or against it. But that also means there's no reason to take it seriously either, for assertions lacking evidence can be dismissed without evidence.

Second, we must construe "proof" not as absolute scientific proof, but in the everyday sense of "evidence so strong you would bet your savings on it." In that sense, we can surely prove that there's no God. This is the same sense, by the way, in which we can "prove" that the earth rotates on its axis, that a normal water molecule has one oxygen and two hydrogen atoms, and that we evolved from other creatures very different from modern humans.

With the notion of a theistic god and a vernacular notion of "proof" in hand, we can disprove a god's existence in this way: *If a thing is claimed to exist, and its existence has consequences, then the absence of those consequences is evidence against the existence of the thing.* In other words, the absence of evidence—*if evidence should be there*—is indeed evidence of absence.

A famous example of this argument is Carl Sagan's chapter "The Dragon in My Garage" in his book *The Demon-Haunted World*. Someone claims that there's a fire-breathing dragon in his garage. The skeptic's demand for evidence is then met with a series of evasions: the dragon is invisible, so you can't see it; it floats, so you can't detect its footprints in scattered flour; its fire isn't hot, so you can't feel its breath. Eventually, says Sagan, the rational course is to reject the dragon's existence until some evidence actually surfaces. His point was that the "you can't prove nonexistence" claim is fatuous when the evidence *should* be there. As he notes at the end of his parable:

Once again, the only sensible approach is tentatively to reject the dragon hypothesis, to be open to future physical data, and to wonder what the cause might be that so many apparently sane and sober people share the same strange delusion.

This was clearly aimed at both pseudoscience and religion, for Sagan was a stronger opponent of faith than most people recall.

We can in fact prove many negatives. Can you prove that I don't have two hearts? Of course you can: just do a CAT scan. Can you prove that I don't have a brother? For all practical purposes, yes: just dig through birth records, ask people, or observe me. You won't find any evidence. Can you prove that I didn't write *Ulysses*? Of course: I wasn't alive when it was published. Can you prove that leprechauns don't live in my garden? Well, not absolutely, but if you never see one, and they have no effects, then you can provisionally conclude that they don't exist. And so it is with all the fanciful features and creatures we firmly believe don't exist.

Many gods claimed to exist *should* have observable effects on the world. The Abrahamic God, in particular, is widely believed to be omnibenevolent, omnipotent, and omniscient. Some also believe that he gives us an afterlife in which we find either eternal bliss or torment, that he answers prayers, and that he had a divine son who can bring us salvation. If these claims are true, there should be evidence for them. But the evidence isn't there: we see no miracles or miracle cures in today's world, much less any wondrous signs of a God who presumably wants us to know him; scientific tests of prayer show that it doesn't work, ancient scriptures show no knowledge of the universe beyond that available to any normal person who was alive when the texts were composed; and science has disproved many of the truth claims of scripture. Finally we are left with that nagging problem of evil: why would a loving and all-powerful God inflict "natural evil" on the world—allowing thousands of innocent people to die from physical disasters like tsunamis, earthquakes, and cancer?

Putting all this together, we see that religion is like Sagan's invisible dragon. The missing evidence for any god is simply too glaring, and the special pleading too unconvincing, to make its existence anything more than a logical possibility. It's reasonable to conclude, provisionally but confidently, that the absence of evidence for God is indeed evidence for his absence.

Science Is Based on Faith

I often hear that science, like religion, is actually based on faith. This argument smacks a bit of desperation, a *tu quoque* response by beleaguered believers. But it also stems from postmodernism's view that even in science, truth is a fungible commodity, with different and incompatible assertions carrying equal weight. As we'll see, the "based on faith" argument against science is purely semantic, resting on two different usages of the word "faith," one religious and the other vernacular.

The surprising thing is that the claim of faith-based science often comes from scientists themselves. Here, for instance, are three religious scientists who argue that accepting the laws of nature is a form of "faith." The first is from the physicist Karl Giberson and the physician and geneticist Francis Collins:

> Finally, we note that it requires a certain level of faith to answer the scientific questions of how something happens. Answers to scientific questions assume that the laws of the universe are constant or, if recent speculations turn out to be true, the laws are changing in only the most subtle ways. This requires faith in the orderliness of nature. With or without belief in an ultimate creator, we must have faith that this univer-

sal order is real, reliable, and accessible to the limited powers of our minds.

The physicist Paul Davies makes a similar claim:

> Clearly, then, both religion and science are founded on faith—namely, on belief in the existence of something outside the universe, like an unexplained God or an unexplained set of physical laws, maybe even a huge ensemble of unseen universes, too. For that reason, both monotheistic religion and orthodox science fail to provide a complete account of physical existence. . . . But until science comes up with a testable theory of the laws of the universe, its claim to be free of faith is manifestly bogus.

Sometimes "faith in science" is meant not just as belief in physical laws, but as blind deference to authority: an unthinking acceptance of the conclusions of scientists in other fields or, if you're a layperson, of scientists in general. This argument was made in, of all places, the pages of *Nature*, one of the world's most prestigious scientific journals. Here Daniel Sarewitz, director of a science and policy think tank, sees belief in the Higgs boson, a particle whose field gives mass to all other particles, as an "act of faith" resembling the superstitions of Hinduism:

> If you find the idea of a cosmic molasses that imparts mass to invisible elementary particles more convincing than a sea of milk that imparts immortality to the Hindu gods, then surely it's not because one image is inherently more credible and more "scientific" than the other. Both images sound a bit ridiculous. But people raised to believe that physicists are more reliable than Hindu priests will prefer molasses to milk. For those who cannot follow the mathematics, belief in the Higgs is an act of faith, not of rationality.

A political science professor at Rutgers University argues that "faith" is often imputed to those of us who rely on Western medicine and its authorities—doctors and medical researchers:

I'm not a biologist; I have never actually seen a microbe in person. But I believe in them. Likewise, I take it on faith when my doctor tells me a particular medication will work in a particular way to address a particular malady.

Finally, the theologian John Haught asserts that the faith of scientists has no philosophical basis: you can't use science itself to show that science is the best way—or even the only way—to discover truths about the universe.

There's the deeper worldview—it's a kind of dogma—that science is the only reliable way to truth. But that itself is a faith statement. It's a deep faith commitment because there's no way you can set up a series of scientific experiments to prove that science is the only reliable guide to truth. It's a creed.

Let's start with the last view, one often raised by philosophers (the argument is called "justificationism"). As a professional scientist, I have always been puzzled by this criticism. It sounds quite sophisticated, and in fact it's technically correct: science cannot justify *by reason alone* that it's the surest route to truth. How can you prove from philosophy and logic alone that scientific investigation, rather than, say, revelation, is the best way to determine the sequence of a newly discovered gene? There's no a priori philosophical justification for using science to understand the universe.

But we don't need one. My response to the "no justification" claim is that the superiority of science at finding objective truth comes not from philosophy but from *experience*. Science gives predictions that work. Everything we know about biology, the cosmos, physics, and chemistry has come through science—not revelation, the arts, or any other "way of knowing." And the practical applications of science, channeled into engineering and medicine, are legion. Many older readers would, like me, be dead were it not for antibiotics, for until these drugs were discovered in the twentieth century, infection was surely the main cause of mortality throughout the evolution of our species. Science has completely eradicated smallpox and rinderpest (a disease of cattle and their wild relatives), is on the way to wiping out malaria and polio, and produced the Green Revolution, saving millions of lives by improving crops and agricultural methods. Every time you use a GPS device, a computer, or a cell phone, you're reaping the benefits of science. In fact, most of us regularly trust our very lives to science: when you have an operation, when you fly in an airplane, when you get your children vaccinated. If you were diagnosed with diabetes, would you go to the doctor or consult a spiritual healer? (I'm appealing to our solipsism here by emphasizing how science has improved human welfare, but most scientists are involved less with helping humanity than with satisfying their own curiosity. After all, our big brains, fueled with food, are still hungry for answers. How old is the universe? How did Earth's species get here? Science alone has given the answers.)

In the end, it may smack of circularity to use empirical results to justify the use of the empirical toolkit we call "science," but I'll pay attention to the circularity argument when someone comes up with a better way to understand nature. Science's results alone justify its usefulness, for it is, hands down, the single best way we've devised to understand the universe. And by the way, if you're going to use the circularity argument against science, you can just as easily apply it to religion. Just as you can't use the Bible as the authority on the divine truth of the Bible, so you can't use philosophy—or any "truth-seeking" method of religion—to show that revelation is a reliable route to the truth.

As for the claim that science is a kind of "faith" because it rests on untestable assumptions, depends on authority, and so on, this involves either a deliberate or an unconscious conflation of what "faith" means in religion versus what it means in everyday life. Here are two examples of each usage:

> "I have faith that because I accept Jesus Christ as my personal savior, I will join my late wife in heaven."
> "I have faith that when I martyr myself for Allah, I'll receive seventy-two virgins in paradise."
> "I have faith that the day will break tomorrow."
> "I have faith that taking this penicillin will cure my urinary tract infection."

Notice the difference. The first two statements exemplify the religious form of "faith," the one Walter Kaufmann defined as "intense, usually confident, belief that is not based on evidence sufficient to command assent from every reasonable person." There is no evidence beyond revelation, authority, and sacred books to support the first two statements. They show confidence that isn't supported by evidence, and most of the world's believers would reject those statements.

In contrast, the second two statements rely on empirical evidence—strong evidence. In these cases the word "faith" doesn't mean "belief without much evidence," but "confidence based on evidence" or "an assumption based on performance." You have "faith" that the Sun will rise tomorrow because it always has, and there's no evidence that the Earth has stopped rotating or the Sun has burned out. You have faith in your doctor because presumably she has treated you successfully in the past and has a good reputation. After all, would you go to a doctor who was constantly being sued for malpractice, or had repeatedly failed to help you? If you had "faith" in your doctor in the religious sense, you'd assume she could do no wrong, no matter what wonky things she'd do or prescribe. If she prescribed toad's blood for your psoriasis, you'd take it gladly. But the kind of faith we *really* have in our doctor is a provisional and evidence-based one—the same kind of "faith" we have in scientific results. After a vigorous but unhelpful regime of toad's blood, you'd find another doctor.

The conflation of faith as "unevidenced belief" with its vernacular use as "confidence based on experience" is simply a word trick used to buttress religion. In fact, you'll almost never hear a scientist using that vernacular in a professional role, saying things like "I have faith in evolution" or "I have faith in electrons." Not only is such language alien to us, but we also know full well how those words can be misappropriated by the faithful.

What about the respect that the public and other scientists have for scientific authorities? Isn't that like religious faith? Not really. When Richard Dawkins talks about evolution and Carolyn Porco about space exploration, scientists in other disciplines accept what they have to say, and the public eagerly consumes their popular books. But that too is based on experience—perhaps not direct experience in the case of the public, but on our understanding that Dawkins's expertise in evolution and Porco's in planetary science have been continuously vetted and accepted by hypercritical scientists.

We know too that the self-correcting nature of science and its tradition of affording more respect to accomplishment than to authority (a common saying is "You're only as good as your last paper") ensure that an incompetent or ham-handed scientist won't gain respect—at least for long. Very few laypeople understand Einstein's theories of relativity, but they know that those theories passed muster with qualified scientists. It was for this reason that Einstein was revered *by the public* as a great physicist. When Daniel Sarewitz claimed that "belief in the Higgs [boson] is an act of faith, not of rationality," and compared it to Hindu belief in a sea of milk, he was simply wrong. There is solid evidence for the existence of the Higgs, evidence confirmed by two independent teams using a giant particle accelerator and rigorous statistical analysis. But there isn't, and never will be, any evidence for a Hindu sea of milk.

In contrast, how reasonable is it to believe that the pope really is infallible when he speaks ex cathedra, or that his views about God are closer to the truth than those of any ordinary priest? A rabbi may gain repute for great kindness or wisdom, but not because he's demonstrated a knowledge of the divine that is more accurate than that of other rabbis. What he may know more about is what other rabbis have *said*. As my friend Dan Barker (a Pentecostal preacher who became an atheist) once quipped, "Theology is a subject without an object. Theologians don't study God—they study what other theologians have said." The claims of a priest, a rabbi, an imam, or a theologian about God have no more veracity than anyone else's. Despite millennia of theological lucubrations, we know nothing more about the divine than we did a thousand years ago. Yes, there are religious authorities, but they aren't equivalent to scientific authorities. Religious authorities are those who know the most about *other* religious authorities. In contrast, scientific authorities are those who are best able to understand nature or produce credible theories about it.

As we've seen, scientists give no special credence or authority to books either, except insofar as they present novel theory, analysis, or data. In contrast, many creeds require believers and ministers to swear adherence to unchanging doctrines like the Nicene Creed, and many Christian colleges have "statements of faith" that must be affirmed yearly by faculty and staff. This distinction, and the fallacy of claiming that science is a religion, was emphasized by Richard Dawkins in an article in the *Humanist:*

There is a very, very important difference between feeling strongly, even passionately, about something because we have thought about and examined the evidence for it on the one hand, and feeling strongly about something because it has been internally revealed to us, or internally revealed to somebody else in history and subsequently hallowed by tradition. There's all the difference in the world between a belief that one is prepared to defend by quoting evidence and logic and a belief that is supported by nothing more than tradition, authority, or revelation.

Scientists, then, don't have faith—in the religious sense—in authorities, books, or unevidenced propositions. Do we have faith in *anything*? Two other objects of scientific faith are said to be physical laws and reason. Doing science, it is said, requires faith not only in the "orderliness of nature" and an "unexplained set of physical laws," but also in the value of reason in determining truth.

Both claims are wrong.

The orderliness of nature—the so-called set of natural laws—is not an assumption but an *observation*. It is logically possible that the speed of light in a vacuum could vary from place to place, and while we'd have to adjust our theories to account for that, or dispense with certain theories altogether, it wouldn't be a disaster. Other "natural laws," like the relative masses of neutrons and protons, probably *can't* be violated, at least in our corner of the universe, because the existence of our bodies depends on those regularities. As I've noted, both the evolution of organisms and the maintenance of our bodies depend on regularities in the biochemical processes that keep all organisms up and running. The laws of nature, then, are regularities (assumptions, if you will) based on experience, the same kind of experience that makes us confident that we'll see another sunrise. After all, Aristotle had "faith" in the religious sense that heavier objects would fall faster than light ones, but it was experiments—sadly, not involving Galileo and the Leaning Tower of Pisa—which showed that, absent air resistance, all objects actually fall at the same rate.

Accommodationists further accuse scientists of having "faith in reason." Yet reason is not an a priori assumption, but a tool that's been shown to work. We don't have faith in reason; we *use* reason, and we use it because it produces results and progressive understanding. Honed by experience to include tools like double-blind studies and multiple, independent reviews of manuscripts submitted for publication, *scientific* reason has produced antibiotics, computers, and our ability to reconstruct the tree of life by sequencing DNA from different species. Indeed, even *discussing* whether we should use reason involves using reason! Reason is simply the way we justify our beliefs, and if you're not using it, whether you're justifying religious or scientific beliefs, you deserve no one's attention.

Another trope in the argument that science is like religion is that we also have a god: the truth revealed by the methods of science. Isn't science, as some maintain, based on a "faith" that it's good to pursue the truth? Hardly. The notion that knowledge is better than ignorance is not a quasi-religious faith, but a *preference:* we prefer to know the truth because accepting what's false doesn't give us useful answers about the universe. You can't cure disease if, like Christian Scientists, you think it's caused by faulty thinking. The accusation that science is based on faith in the value of knowledge is curious, for it's not applied to other areas. We don't argue, for instance, that plumbing and auto mechanics are like religion because they rest on an unjustified faith that it's better to have your pipes and cars in working order.

Religion Gave Rise to Science

Even if you can't show harmony between science and religion, you can always argue that science was a *product* of religion: that, long ago in Europe, modern science arose from religious beliefs and institutions. Given that science as practiced now is completely free from gods, this is a strange argument, but it's a way to give religion, even if incompatible with modern science, some credit for that science. And given the preponderance of Western theists who make this argument, it's no surprise that it's Christianity rather than Judaism or Islam that gets the credit.

This argument takes several forms. The most common is that science came from natural theology, which itself arose from the Christian desire to understand God's creation. The most detailed version of this argument comes from the sociologist Rodney Stark:

The rise of science was not an extension of classical learning. It was the natural outgrowth of Christian doctrine: nature exists because it was created by God. In order to love and honor God, it was necessary to fully appreciate the wonders of his handiwork. Because God is perfect, his handiwork functions in accord with immutable principles. By the full use of our God-given powers of reason and observation, it ought to be possible to discover these principles.

Almost as common is the claim, made here by Paul Davies, that the concept of physical law itself came from Christianity:

> The very notion of physical law is a theological one in the first place, a fact that makes many scientists squirm. Isaac Newton first got the idea of absolute, universal, perfect, immutable laws from the Christian doctrine that God created the world and ordered it in a rational way. Christians envisage God as upholding the natural order from beyond the universe, while physicists think of their laws as inhabiting an abstract transcendent realm of perfect mathematical relationships.

As we'll see, these claims are disputed, but even if they're wrong, theists can always fall back on the argument that the *ethics* undergirding modern science come from Christian morality. As Ian Hutchinson argues, "The ethical and moral acceptability of scientific practices is strongly dictated by religious beliefs and commitments."

To address the Christianity-produced-science argument, we should realize that science arose in other places before Christian Europe, most notably ancient Greece, the Islamic Middle East, and ancient China. But because *modern* science is essentially a European invention whose spirit and motivations derived from ancient Greece and Rome, various explanations are given for why it fizzled out elsewhere. Islamic science, for instance, is often said to have disappeared after the twelfth century because free inquiry was declared inimical to Quranic doctrine. But explaining such large-scale social change is often slippery, susceptible to multiple and conflicting interpretations. Some Christian apologists, like the mathematician Alfred North

Whitehead, argue that faith in the "order of nature" and "general principles" (i.e., physical laws) was inherent in medieval Christianity:

> My explanation [for why science developed in Europe and not other areas] is that faith in the possibility of science, generated antecedently to the development of modern scientific theory, is an unconscious derivative from medieval theology.

But one can argue even more cogently that the idea that the universe could be understood through reason was a legacy of ancient Greece.

Another strategy is to argue, as does Ian Hutchinson, that many famous scientists were religious, and their work was motivated by their faith:

> Any list of the giants of physical science would include Copernicus, Galileo, Kepler, Boyle, Pascal, Newton, Faraday, Maxwell, all of whom, despite denominational and doctrinal differences among them, and opposition that some experienced from church authorities, were deeply committed to Jesus Christ.

This is related to the claim that science and religion are *compatible* because many scientists are still religious.

What can we make of these claims? It would be petulant to argue that religion, or Christianity in particular, made no contribution to science, or has always impeded science. Some scientists, like Newton and the nineteenth-century British natural theologians, apparently *were* motivated by their faith, and produced valuable work as a result. Some medieval theologians argued that God gave us reason to help us to understand the world. Monasteries were often the only repositories of scientific knowledge from earlier thinkers. And churches helped create and support European universities in the Middle Ages, some of which encouraged prescientific inquiry.

Overall, however, the assertion that "religion birthed science" doesn't hold water. But first we must admit that even if this thesis were true, it gives no credence to the tenets of faith, or to the value of religion in finding truth. Even institutions founded on falsity can sometimes mature by casting aside their childish things. Alchemy was the predecessor of chemistry—Robert Boyle, who made immense contributions to chemistry, also dabbled extensively in alchemy—but we've long since abandoned the notion of turning lead into gold. Boyle's accomplishments in chemistry don't burnish the image of alchemy.

And if Christianity was *required* for science to emerge, why was there such a burst of science in ancient Greece and Rome, as well as China and Islamic countries? Many ancient Greeks and Romans embraced rationalism and scientific inquiry as a way to understand the world. Think of the accomplishments of people like Aristotle, Ptolemy, Pythagoras, Democritus, Archimedes, Pliny the Elder, Theophrastus, Galen, and Euclid. As the historian Richard Carrier has argued, if any faith should get credit for science, it would be paganism. And there's little evidence that Greek and Roman science was anything other than a secular endeavor motivated by pure curiosity.

The historians Richard Carrier, Toby Huff, Charles Freeman, and Andrew Bernstein have noted that although Christianity took hold in Europe about 500 CE, science didn't come into its own until much later. In their view (which is, of course, contested), the authoritarianism of the church suppressed the kind of freethinking that really did produce modern European science. Heresies like Arianism (the notion of God not as a trinity but a single being) and Manichaeism (the belief that God is benevolent but not omnipotent) were brutally suppressed. Indeed, the notion of "heresy" itself is explicitly antiscientific. If science required Christianity for its genesis, why that thousand-year delay? Why, if Christianity promoted scientific innovation during the Middle Ages, did Europe show no economic growth for a millennium? Reviewing Rodney Stark's defense of Christianity as critical for the birth of science, Andrew Bernstein argued that the hiatus of science during the Dark Ages reflected the diversion of brainpower from empirical issues to apologetics:

> In the Middle Ages, the great minds capable of transforming the world did not study the world; and so, for most of a millennium, as human beings screamed in agony—decaying from starvation, eaten by leprosy and

plague, dying in droves in their twenties—the men of the mind, who could have provided their earthly salvation, abandoned them for otherworldly fantasies. Again, these fundamental philosophical points bear heavily against Stark's argument, yet he simply ignores them.

The notion that Christianity was pivotal in producing science also fails to explain why science didn't arise in the *Eastern* Empire, which was Christian, prosperous, and endowed with rich libraries holding the scientific works of ancient Greeks and Romans.

In the end, we don't know why modern science arose for keeps in Europe between the thirteenth and sixteenth centuries, while arising and then vanishing in China and Islamic countries. Besides Christianity, there were other differences between the West and other areas that could have promoted European science, including the advent of the printing press, the greater mobility of Europeans, a critical mass of population that could promote intellectual interaction, and the questioning of authority (including the religious kind)—in other words, everything that brought about the Enlightenment. The rise of modern science in Europe is a complex affair that, as a one-off historical event, defies conclusive explanation. Christianity may be one factor, but we can't rerun the tape of history to see if science would have arisen later in a Europe that lacked religion.

But we can at least show that, in some respects, Christianity impeded free inquiry. Many theologians from Aquinas on advocated the killing of heretics, hardly an endorsement of freethinking. Martin Luther was famous for his attacks on reason. Besides persecuting Galileo and Giordano Bruno for their heresies, some of which involved science, and burning Bruno alive, the Catholic Church famously condemned the University of Paris in 1277 for teaching 219 philosophical, theological, and scientific "errors." And what are we to make of the church's infamous *Index Librorum Prohibitorum*, which for four centuries protected its flock from theologically incorrect thinking? That apparently included science, for the list included works by Kepler, Francis Bacon, Erasmus Darwin (Charles's grandfather, who had his own theory of evolution), Copernicus, and Galileo. Why would an institution that promoted science make it a sin to read books by scientists? And

why would an institution in favor of free inquiry ban philosophy books by
Pascal, Hobbes, Spinoza, and Hume?

Finally, what about those famous scientists who were religious? We
shouldn't be too quick to give scientific credit to their Christianity. Newton,
for instance, was an Arian who rejected the Trinity, the divinity of Jesus, and
the notion of an immortal soul. But the argument that the existence of
Christian scientists proves that Christianity caused science is wholly unconvincing, for it's based simply on correlation. In medieval and Renaissance
Europe, nearly *everyone* was a Christian, or at least professed to be, simply
because it was a universal belief that prominent people defied at peril of execution. If Christianity gave rise to science between the twelfth and sixteenth
centuries, then you could give religion credit for *everything* that humans devised in that period.

And we can firmly reject any contribution of religion to *modern* science.
As we know, scientists are on average far less religious than are laypeople,
and the most accomplished scientists are nearly all atheists. This means
that virtually no modern scientific research *can be* motivated by religion,
and I'm aware of no scientific advances made by those who claimed religious inspiration. Most of the major scientific achievements of our time—
advances in evolution, relativity, particle physics, cosmology, chemistry, and
modern molecular biology—were made by nonbelievers. (While intelligent
design creationism does have religious roots, it is those very roots that have
discredited it as valid science, for there's simply no evidence for the claimed
intervention of a teleological designer in evolution.)

James D. Watson once told me that while searching for the structure of
DNA, he and Francis Crick were strongly motivated by naturalism: they
wanted to show that the "secret of life"—the replicating molecule that is the
recipe for all organisms—was pure chemistry, with no divine intervention
required. If we're going to give religion credit for the birth of science, then
by the same lights we must give nonbelief credit for most of the scientific
advances of the last century, which were driven by ruthless adherence to
naturalism. Every bit of truth clawed from nature over the last four centuries has involved completely ignoring God, for even religious scientists park
their faith at the laboratory door.

As for religion's positive contribution to the *morality* of science, the case is weak. You'd be hard pressed to show that the ethics imbuing modern science—treating laboratory animals humanely, not falsifying data, giving people due credit for their contributions—come from religious beliefs rather than secular reason. And religious morality has clearly impeded modern science, by producing bans on much stem cell research, promoting the AIDS epidemic through Catholic claims that condoms don't prevent the disease (as well as encouraging population growth by discouraging contraception), and hindering vaccination through religiously based opposition by Muslims and Hindus.

Religion has undoubtedly contributed to the work of some scientists, and may even have played some role in the rise of the discipline, at least through sponsoring universities that nurtured early scientists. But balancing religion's beneficial versus repressive role in science is a task for historians, who, after much bickering, have failed to reach a consensus.

Science Does Bad Things

Defenders of religion often try to balance the undeniable benefits of science by arguing that it has also been responsible for many of the world's woes. The biologist Kenneth Miller's argument is typical:

> Science is a revolutionary activity. It alters our view of nature, and often puts forward profoundly unsettling truths that threaten the status quo. As a result, time and time again, those who feel threatened by the scientific enterprise have tried to restrict, reject, or block the work of science. Sometimes, they have good reason to fear the fruits of science, unrestrained. To be sure, it was religious fervor that led Giordano Bruno to be burned at the stake for his scientific "heresies" in 1600. But we should also remember more recently that it was science, not religion, that gave us eugenics, the atomic bomb, and the Tuskegee syphilis experiments.

A similar note is sounded by the novelist Jeffrey Small, an Episcopalian:

> Critics of religion enjoy pointing out how many wars and how much suffering has been caused in the name of religion. But only science has given

us the tools to kill each other in ways never before imagined. Biologists have produced viral and bacterial weapons; chemists have developed gunpowder and ever more destructive explosives; physicists have given us the power to destroy our very existence with nuclear weapons. Scientific advances in mechanical and chemical engineering have made our businesses more productive than at any time in history, bringing us comfort and prosperity. These same advances have also polluted our environment to the point of endangering our planet.

Statements like these have one aim: to show that while religion may have done bad things, science has too. They are *tu quoque* arguments: "See, you're as bad as we are!" As the journalist Nick Cohen noted about accusations that atheism is like religion, "It's not a charge I'd throw around if I were seeking to defend faith. When people say of dozens of political and cultural movements from monetarism to Marxism that their followers treat their cause 'like a religion,' they never mean it as a compliment. They mean that dumb obedience to higher authority and an obstinate attachment to dogma mark its adherents."

Notice that these indictments are aimed not at scientists, but at "science"—as if the discipline itself, rather than its practitioners, is responsible for this malfeasance. But science is simply a way of investigating the world, a set of tools to discover what's out there. The compelling force that produced nuclear weapons, gunpowder, and eugenics was not science but *people:* the scientists who decide to use discoveries in a certain way, the technologists who convert those discoveries into things like weapons, and the people who make decisions to use technology for purposes that may be harmful or immoral. Although physicists produced the work on nuclear fission that made possible the atomic bombs dropped on Japan, the executive order that started work on the bomb in America was signed by Franklin Roosevelt, and the Manhattan Project was directed not by a scientist, but by a soldier, Major General Leslie Groves. Roosevelt's decision was in fact partly a response to a letter he received from Albert Einstein urging the United States to stockpile uranium because Einstein feared (rightly) that Germany was trying to develop an atomic bomb. The decision to drop the bomb on Japan was made by President Harry Truman. In other words, between the science itself and its devastating effects was a chain of people making tactical and ethical decisions.

The findings of science are morally neutral; it is how they are *used* that is sometimes a problem. While one might be tempted to make a similar argument about religion, I'll claim that there are important differences between science and faith that make religion itself complicit in its misuse.

When I read indictments of science for its harmful results, I think of the following: "Toolmaking has given us shovels, hammers, chisels, and knives. But sometimes those tools are used to kill people, so we must remember that, although a valuable enterprise, toolmaking has also brought us misery." But, like those of science, the misuses of toolmaking are far outweighed by its benefits. Blaming a field of endeavor, rather than misguided people, for its misuse is like blaming architecture for giving Nazis the means to build gas chambers. And when you pin overpopulation and pollution on "science," is it really the institution and methodology that are to blame, or is it greedy, shortsighted people? Are Darwin and Mendel to blame for eugenics, or is it the corruption of those enterprises by racists and xenophobes? In the end, the solution is not to stop science, or even blame science, but to correct the mind-set that results in bending it toward nefarious or socially harmful ends. Clearly, so long as science is practiced by humans it will never be free from misuse by bad people. And it will always have some bad effects.

But then what about religion? If we can exculpate science for the ills it causes, can't you exculpate religion on the same grounds? Can't you say that the evils of religion—things like the Inquisition or the terror bombings of radical Muslims—come not from dogma or scripture but from their misapplication by flawed human beings? My answer is that here science differs from religion in an important way: unlike science, *faith itself can corrupt decent people,* leading directly to bad behavior.

Most religions, and certainly the Abrahamic ones, have three features that are foreign to science. The most important is religion's linkage to moral codes that define and enforce proper behavior, behavior supposedly reflecting God's will. The second is the widespread belief in eternal reward and punishment: the notion that after death not just your fate but everyone else's depends on adherence to conduct mandated by *your* religion. And the third is the notion of absolute truth: that the nature of your god, and what it wants, is unchanging. While some believers see their ability to fathom

God's nature as limited, and don't accept the notion of a heaven or a hell, the certainty of religious dogma is far more absolute and far less provisional than the pronouncements of science.

This combination of certainty, morality, and universal punishment is toxic. It is what leads many believers not only to accept unenlightened views, like the disenfranchisement of women and gays, opposition to birth control, and intrusions into people's private sex lives, but also to force those views on others, including their own children and society at large, and sometimes even to kill those who disagree. It is this toxic mixture, which we'll discuss in the next chapter, that the physicist Steven Weinberg indicted when he said, "With or without religion, good people can behave well and bad people can do evil; but for good people to do evil—that takes religion." He did not mean, of course, that religion turns all good people bad, but merely some of them, depending on their religion and their ardor. Without religion, for instance, it's hard to imagine the eternal enmity between Sunni and Shiite Muslims, often people from identical national and ethnic backgrounds who nevertheless slaughter each other over the question of who were Muhammad's proper heirs. The eternal persecution of the Jews is a purely religious matter, turning on their presumed status as killers of Christ.

But the same things cannot be said of science, for the discipline contains nothing prescriptive (save "find the truth" and "don't cheat"), nor any intimation of eternal rewards. Physicists do not kill each other when they differ about the value of string theory or who first came up with the idea of evolution.

Actually, Weinberg wasn't quite correct. For good people to do evil doesn't require only religion, or even *any* religion, but simply one of its key elements: belief without evidence—in other words, faith. And that kind of faith is seen not just in religion, but in any authoritarian ideology that puts dogma above truth and frowns on dissent. This was precisely the case in the totalitarian regimes of Maoist China and Stalinist Russia, whose excesses are often (and wrongly) blamed on atheism. And it is in such societies, where free inquiry is suppressed, that we find bad science becoming an institution.

Perhaps the most famous example of pernicious, ideology-based science is the "Lysenko affair," in which a bogus form of genetics held sway in the

Soviet Union between 1935 and the mid-1960s. "Lysenkoism" was a cult of personality centered on both Stalin and his handpicked "expert" in agriculture, the mediocre agronomist Trofim Denisovich Lysenko. Catching Stalin's ear with exorbitant and bogus claims that he could produce more crops by treating seeds with extreme cold and moisture, Lysenko became in effect the dictator of Soviet agriculture and genetics. His methods rested on the unscientific and unsubstantiated claim that environmental treatments could affect the heredity of plants, a claim that conflicts with everything we know about genetics. Western genetics and plant breeding were abandoned as decadent, and, with Lysenko's approval, famous geneticists were either executed or sent to the gulag. Other scientists, hoping to avoid punishment, simply faked their data to conform to Lysenko's ideas.

Lysenkoism failed miserably. It didn't improve crop yield, and the purge of geneticists set Soviet biology back by decades. Is it then a black mark on science? Hardly, for it marked the abandonment of real science for something like creationism: empirical statements based on wish-thinking and supported by fealty to a religious-like god (Stalin) and his anointed son (Lysenko). It was the faith in these methods, and the suppression of the normal criticism and dissent of science, that caused the debacle. As Richard Feynman said in his report on the failed O-rings that doomed the space shuttle *Challenger*, "For a successful technology, reality must take precedence over public relations, for nature cannot be fooled."

But Weinberg *was* on the money when he (and the philosopher Karl Popper) argued that the problems imputed to "science" are really the problems afflicting all of humanity: venality, irrationality, and immorality:

Of course science has made its own contribution to the world's sorrows, but generally by giving us the means of killing each other, not the motives. Where the authority of science has been used to justify horrors, it really has been in terms of perversions of science, like Nazi racism and "eugenics." As Karl Popper has said, "On the other hand, it is only too obvious that it is irrationalism and not rationalism that has the responsibility for all national hostility and aggression, both before and after the Crusades, but I do not know of any war waged for a 'scientific' aim, and inspired by scientists."

Science Is Fallible and Its Results Are Unreliable

This is another *tu quoque* argument from beleaguered believers. If religion can be wrong, the argument runs, then so can science. If we have our doubts about religious truths, well, scientific truths are also shaky. After all, hasn't scientific "knowledge" been overturned time after time? The author Jeffrey Small expresses this sentiment in an article called "The Common Ground Between Science and Religion":

> We must also be careful not to overstate the infallibility of the scientific method. Scientific knowledge has inherent limitations. Science is not truth; it's an approximation of truth. . . . Another limitation with the scientific method is that all scientific theories rely on human conception, interpretation and evaluation. The history of science shows that the process of one scientific theory supplanting another is a bumpy one.

This argument is not unique to religionists and accommodationists: it's a staple of postmodernists and assorted pseudoscientists, including advocates of creationism, alternative medicine, global-warming denialism, and the supposed dangers of vaccination. In Texas, for instance, the "science is wrong" trope appears in the biology curriculum of a publicly funded "charter school":

> Many other historical blunders of science could be mentioned. What we need to keep in mind is that scientists are human beings. The assumption that they are completely objective, error-free, impartial, "cold machines" dressed in white coats is, of course, absurd. Like everyone else, scientists are influenced by prejudice and preconceived ideas. You should also remember that just because most people believe a particular thing does not necessarily make it true.

The "particular thing" under discussion is, of course, evolution.

The response is simple. *Of course* science can be wrong, and has been many times before—but that's what's right about it. Naturally scientists are only human, and sometimes reluctant to part with their pet theories, but they also make mistakes. That, in combination with the limited understanding we have at any one time, guarantees that many scientific "truths" will fall by the wayside. Some scientific results are flat wrong, as in the cases of faster-than-light neutrinos, cold fusion, bacteria with arsenic in their DNA, and the notion of static continents, while others have simply replaced useful paradigms, like Newtonian mechanics, with more inclusive ones, like quantum mechanics. And beyond simple error, there's been fraud. The most famous case is Piltdown Man, a hoax involving a humanlike skull. Supposedly found in a gravel pit in East Sussex, the skull was shown to scientists in 1912 by the amateur archaeologist Charles Dawson, stunning them with its combination of a modern skull and apelike teeth. Many saw it as a transitional form between primitive and modern humans: proof that we evolved. It took four decades for that skull to be revealed as a forgery, a mélange of a medieval human skull, the jaw of an orangutan, and teeth from a chimpanzee. (The identity of the forgers remains a mystery, but suspects include the writer Arthur Conan Doyle and the Jesuit priest and paleontologist Pierre Teilhard de Chardin.) Piltdown Man is still a staple of creationist literature, regularly trotted out to show that evolutionary biology and its practitioners can't be trusted.

But notice that all of these hoaxes and false results were exposed *by scientists themselves.* It was suspicious anthropologists and paleontologists who uncovered the Piltdown forgery, so there was no collusion (as implied by creationists) to buttress the "lie" of evolution with a phony fossil. And, of course, we now have a panoply of genuine fossils attesting to human evolution. The "arsenic" DNA, faster-than-light neutrinos, and cold fusion were all quickly debunked by other scientists trying to replicate the results.

And that's the point. Science has a huge advantage over "other ways of knowing": built-in methods of self-correction. These include not only the familiar attitude of doubt, but also an arsenal of empirical weapons to test and replicate the results of others. After all, renown accrues to scientists who show up their peers (we're just as ambitious as anyone else), and one way to do that is to disprove a result that has gained a lot of attention. Certainly some scientists are reluctant to part with theories that have made them famous, and paradigms do get entrenched in some fields (the idea that continents don't move is one example), for scientists are, with good reason, conservative. But ambition and the desire to know will ultimately lead to good science driving out the bad.

Although scientific research may change some of our conclusions over the years, many of those conclusions will remain intact. It's unlikely, for instance, that we'll find that continental drift is wrong, for we can actually see and measure the movement of continents using satellites and lasers. Few scientists doubt that, several centuries from now, DNA will remain the genetic material in multicellular species, that the speed of light in a vacuum will remain within 1 percent of its reported value, and that a molecule of methane will have one carbon and four hydrogen atoms. These things can be regarded, in the vernacular, as "proven." In fact, we've seen that the very people who argue that science is fallible and its results are untrustworthy put their trust in science every day. Why would they do that?

But none of this criticism of science makes religion even a tiny bit more credible. While science has been wrong, it's been right enough to improve our understanding of the universe in a way that's immeasurably advanced the well-being of our own species and our understanding of nature. Even a simple scientific advance can save millions of lives. The Green Revolution is well known, but a more recent innovation is the development of "golden rice," a genetically engineered crop that incorporates a precursor to vitamin A, an essential nutrient, into the rice genome. The product is nutritious, perfectly safe, distributed without charge to subsistence farmers, and, best of all, could save the lives of nearly three million children who die annually of vitamin deficiency. Sadly, misguided people who are suspicious of all genetically modified organisms—GMOs—have prevented widespread distribution of the product.

In contrast, religion has *never* been right in its claims about the universe—at least not in a way that all rational people can accept. There is no reliable method to show that the Trinity exists, that God is loving and all-powerful, that we'll meet our dead relatives in the afterlife, or that Brahma created the universe from a golden egg. Lacking a way to show its tenets are wrong, religion cannot show them to be right, even provisionally.

Although this chapter may have had the flavor of an academic debate, what with the emphasis on charge and countercharge, argument and answer, the stakes are far higher than simple intellectual victory. For mixing science with faith, or assuming that they are coequal ways of finding truth, harms not just intellectual discourse but also people's lives. The next chapter describes the damage of such accommodationism.

CHAPTER 5

Why Does It Matter?

A surgeon once called upon a poor cripple and kindly offered to render him any assistance in his power. The surgeon began to discourse very learnedly upon the nature and origin of disease; of the curative properties of certain medicines; of the advantages of exercise, air and light, and of the various ways in which health and strength could be restored. These remarks were so full of good sense, and discovered so much profound thought and accurate knowledge, that the cripple, becoming thoroughly alarmed, cried out, "Do not, I pray you, take away my crutches. They are my only support, and without them I should be miserable indeed!" "I am not going," said the surgeon, "to take away your crutches. I am going to cure you, and then you will throw the crutches away yourself."

—Robert Green Ingersoll

Even if you agree that science and religion are incompatible, what's the harm in that? After all, most religions aren't opposed to science in general, and many religious scientists happily ignore God while they do their day job, even if they abandon that attitude when they go to church on Sunday.

The harm, as I've said repeatedly, comes not from the existence of religion itself, but from its reliance on and glorification of *faith*—belief, or, if you will, "trust" or "confidence"—*without supporting evidence*. And faith, as employed in religion (and in most other areas), is a danger to both science and society. The danger to science is in how faith warps the public understanding of science: by arguing, for instance, that science is based just as strongly on faith as is religion; by claiming that revelation or the guidance of ancient books is just as reliable a guide to truth about our universe as are the tools of science; by thinking that an adequate explanation can be based on what is personally appealing rather than on what stands the test of empirical study.

Religious scientists undermine their own profession by diluting the rigor of science with claims about the supernatural—claims that are, broadly construed, scientific. Despite Stephen Jay Gould's declaration that "proper" religion stays away from making assertions about the natural world (we've learned that, for theistic religions, there is no clear distinction between the "natural" and the "supernatural" world), religion regularly becomes improper, making clear claims about reality. Both scientists and theologians have shown that Gould was wrong in asserting that Abrahamic religions are, in Ian Hutchinson's words, "empty of any claims to historical or scientific fact, doctrinal authority, and supernatural experience."

Historical facts are, of course, scientific facts, but the new natural theology also makes scientific *claims*. Perhaps the most damaging are the "god of the gaps" arguments: caulking the holes in our understanding of nature with divine explanations. Not only are such explanations easily destroyed by the advances of science (and this has happened repeatedly), but they also give people the false impression that some questions about the universe are simply refractory to scientific explanation, for the explanation lies outside of science.

When Francis Collins argues that because altruism and innate moral feelings cannot be explained by science, and therefore must have been given to us by God, he's making a claim about both nature and science: morality will *always* elude naturalistic explanations. When theologians argue that both consciousness and the ability of our senses to detect truth will never be explained by science, they are not only misleading the public, but acting as "science stoppers," implicitly suggesting that scientists should simply give up studying these phenomena. When religious biologists say that the evolution of humans, or of a humanlike species, was inevitable, they are making a claim that *sounds* scientific, but really rests on scriptural notions of humans as God's special species. When theistic evolutionists state that God acts by moving electrons around in an undetectable way, or by creating the odd mutation to produce a desired species, they are making claims that have no scientific basis but are superfluities tacked on to science to fulfill emotional needs. And when we hear that the laws of physics, and the so-called fine-tuning of the universe's physical constants, have no explanation save God, I know that nearly all physicists disagree.

But does the public *hear* their disagreement and understand their counter-arguments? More likely the layperson, at least in the United States, thinks that, yes, science has indeed reached its explanatory limits, and beyond those limits lies God. That is a distortion of science. As we've learned, science does indeed have provisional explanations for morality, altruism, consciousness, the specificity of the laws of physics in our universe, and the fact that many of our beliefs are true. Those explanations may be wrong, but how can we know without even more science? Unfortunately, "god of the gaps" arguments discourage further research by claiming that science *can never produce* such explanations.

As a scientist, I am distressed by this constant elbowing of religion into questions of reality, and even more so when it leads to unsubstantiated claims about evolution. As we've seen, religion has no warrant and no method for decreeing what is and what is not beyond science. Certainly science has some hard problems, and just as certainly some of those problems will never be solved (why *is* the speed of light in a vacuum constant?), simply because the final answer will be "That's just the way it is." We may reach the limits of explanation for several reasons: because the evidence eludes us (many ancient species, for example, simply weren't fossilized) or because our brains aren't configured to puzzle out the answers. But consider how many questions religion once told us could never be answered—and were taken as evidence for God—and yet ultimately were solved by science. Evolution, infectious disease, mental illness, lightning, the stable orbits of planets: the list is long. Religious people often call for scientists to be "humble," ignoring the beam in their own eyes, which see things like morality as forever inexplicable by science. How much more arrogant, and ignorant of history, to argue that our failures of understanding are somehow evidence for a god! And how much more egotistical to believe that that god is the god of *your own religion*! If the "other ways of knowing" of your faith provide concrete answers, then tell us not only what those answers are, but how they would convince either nonbelievers or members of other faiths. And let those "other ways of knowing" make predictions in the same way that science does.

The damage to science I've emphasized so far involves the public perception of science. The practice of science itself isn't seriously harmed by accommodationism, but there is one exception. And that is when the direction of science is warped by organizations like the John Templeton Foundation, which can actually steer research down certain avenues congenial to its aims: the harmony among science, faith, and spirituality. Not all of Templeton's funding goes to that kind of research, but one can argue that because of its priorities there is more work on "spiritual" topics like near-death experiences than we'd have if scientists themselves (as they do in many government agencies) decide which research gets funded. The "core funding areas" of the Templeton Foundation in the life sciences include these:

> The Foundation supports projects investigating the evolution and fundamental nature of life, human life, and mind, especially as they relate to issues of meaning and purpose. Projects are welcome from a variety of disciplinary perspectives, including the biological sciences, neuroscience, archeology, and paleontology.

"Meaning and purpose" are human constructs, products of intelligent minds, and "purpose" implies forethought of such minds, either human or divine. These are teleological ideas that are not part of science, except in work on human behavior. Here we see the subtle bending of scientific research toward unanswerable religious questions.

We see similar distortion in Templeton's funding of the human sciences:

> The Foundation supports projects that apply the tools of anthropology, sociology, political science, and psychology to the various moral and spiritual concepts identified by Sir John Templeton. These include altruism, creativity, free will, generosity, gratitude, intellect, love, prayer, and purpose.

Clearly these areas are motivated by curiosity not about nature but about the numinous.

Child Abuse: Faith as Substitute for Medicine

But far worse things happen when faith, seen as a valid route to empirical truth, is accompanied by other aspects of religion: the notions that you possess absolute truth about divine aspects of the universe, that adherents to other faiths are simply wrong, and that God has given you a code of behavior enforced by a system of eternal rewards and punishments. That can lead to missionizing: attempts to enforce one's unsubstantiated beliefs on others. And while more liberal religions avoid such missionizing (have you ever had a pair of Unitarians knock on your door?), they often act as enablers of more extreme, antiscience creeds. If religious faith is generally harmful, as I think it is, then any religion whose beliefs rest on faith or that extol faith contributes to that harm.

Nowhere is this missionizing, and its support by religion in general, more toxic than in those sects that reject medical care in favor of prayer and faith healing, and enforce this belief on their children. Denied the benefits of modern scientific medicine, those children often endure prolonged and horrible deaths. Their stories are appalling testimony not only to the incompatibility of science and faith, but to the fact that this incompatibility is embraced not just by biblical literalists, but by members of more sophisticated and less marginalized faiths. And all of us, even nonbelievers, have contributed to these deaths, at least in the United States, by passing laws allowing children to be denied medical care on religious grounds. Underlying it all is the privileging of faith—giving a pass to religious beliefs that contradict science.

Christian Science (the official name is the Church of Christ, Scientist) is not just an oxymoron, but also a mainstream faith, with over a thousand churches in the United States and perhaps several hundred thousand members worldwide (the numbers are kept secret). Its members are not Bible-thumping fundamentalists, but often educated and affluent members of the community. Because Christian Scientists believe that disease and injury are illusions caused by faulty thinking, many of them reject modern medicine, relying instead on Christian Science "practitioners" who are given a mere two weeks of training—none of it in genuine medical care. The church also runs sanatoriums and nursing homes where patients are given prayer instead of medicine. Curiously, Christian Scientists are allowed to go to dentists and optometrists—apparently bad teeth and eyes are exceptions to the view of bodily infirmities as illusions—and can have broken bones set. Many of them also supplement Christian Science "healing" with modern medicine, though that's against church rules. But when they treat their children's maladies with prayer alone, the results are heartbreaking, for the children are either too young to understand or have been indoctrinated into the dogma of faith healing. One of the most horrible cases involved a young girl, Ashley Elizabeth King.

Ashley was the only child of Catherine and John King, prosperous middle-class Christian Scientists in Phoenix (John was a real estate developer). In 1987, at the age of twelve, Ashley developed a lump on her leg. Her parents sought no medical aid, and the lump continued to grow. When it became too large and painful to allow her to go to school, they withdrew her, and although Ashley was supposed to receive in-home instruction by teachers, her parents refused it.

Ashley's lump—a tumor—kept growing, and the Kings continued to ignore it. In May 1988, a detective, alerted by neighbors who hadn't seen the child for months, managed to enter the Kings' home, and saw that the problem was serious. Although Ashley tried to cover the tumor with a pillow, the detective immediately realized that she was in fact dying. A court order put her in custody of child protective services, which sent her to Phoenix Children's Hospital. But by the time she got real medical attention, it was far too late. Her tumor was an osteogenic sarcoma—bone cancer—and had metastasized to her lungs. Her heart was dangerously enlarged from trying to pump blood to the growing tumor, and since she couldn't move because of the pain, her genitals and buttocks were covered with bedsores. Her tumor had grown to thirteen inches across, larger than a basketball, and the stench from her rotting flesh permeated the hospital floor. The doctors recommended amputating the leg—not to save her life, for her condition was terminal—but to ease her pain and give her a bit more time. One doctor said that Ashley was experiencing "one of the worst kinds of pain known to mankind."

The Kings refused amputation, and on May 12 moved their daughter to a Christian Science sanatorium where there was no medical care, not even pain medication. Instead, there were seventy-one calls to Christian Science practitioners for Ashley's "treatment": prayer alone. When she cried out in agony, she was told that she was disturbing the other patients. Ashley died on June 5, 1988, a martyr to her parents' delusions. At the subsequent trial of her parents, a prosecutor described her tumor at death as "about the size of two watermelons." The doctors believed that had she been diagnosed early, there was a 50 to 60 percent chance she could have been saved.

Arizona is one of the few states that don't give parents immunity from prosecution for child abuse if they withhold medical care on religious grounds. (If you withhold it on nonreligious grounds, you're culpable everywhere.) The Kings were tried for that abuse after a charge of negligent homicide was dropped. They pleaded no contest, were convicted of one count of reckless endangerment, a misdemeanor, and were given a slap on the wrist—three years' *unsupervised* probation and 100 to 150 hours of community service.

As in many such cases, the parents showed a curious lack of affect and remorse for what they had done. In a press conference after Ashley's death, her mother compared her daughter's fear of being hospitalized to Anne Frank's anguish about her deportation to Auschwitz. Catherine King added, "I know I was a good mother, and no judge or jury in the country can convince me otherwise." In contrast, the county attorney who filed charges against the Kings said, "Any person who calls himself a Christian wouldn't let a dog die like this." But it was precisely *because* the Kings were members of a Christian sect, one with beliefs about healing, that Ashley died in misery. Had the Kings been atheists, there was a good chance she would have lived.

I've read dozens of these cases, and their common elements are two: no serious punishment of the parents, and those parents' lack of regret. Both, I think, are due to faith. Most states have no legal grounds to prosecute parents like Ashley's; in those where they can be put on trial, juries are reluctant to convict and judges reluctant to punish. But that occurs only when the abuse has a religious excuse. I attribute the lack of affect in the parents to religion as well: the belief that the pain, suffering, and death of their children are far less important than not violating the tenets of their faith. Their conviction that they did what God or their church demands immunizes such parents against normal feelings of guilt and shame.

While there are reams of similar stories about the deaths of children after Christian Science "treatment," many other, more marginal sects also reject medical care in favor of prayer. This dogma invariably rests on passages in the Bible such as James 5:13–15:

> Is any among you afflicted? let him pray. Is any merry? let him sing psalms. Is any sick among you? let him call for the elders of the church; and let them pray over him, anointing him with oil in the name of the Lord: And the prayer of faith shall save the sick, and the Lord shall raise him up; and if he have committed sins, they shall be forgiven him.

Medical advances have been incalculably large since those words were written, but our children continue to suffer and die on the basis of ancient texts—and from the peer pressure exerted by coreligionists. The many sects that rely on prayer treatment often shun or expel members caught going to doctors.

Jehovah's Witnesses, numbering nearly eight million worldwide, routinely refuse blood transfusion, citing biblical passages like Genesis 9:4 ("But flesh with the life thereof, which is the blood thereof, shall ye not eat") and Leviticus 17:10 ("I will even set my face against that soul who eateth blood, and will cut him off from among his people"). Many adults and children have died from a metaphorical interpretation of "eating blood," although transfusing some *components* of blood, like the protein hemoglobin, is now permitted. The children who have died, brainwashed by their parents into refusing blood, are celebrated by Jehovah's Witnesses as martyrs: a copy of the church's magazine *Awake!* from May 1994 shows pictures of twenty-five of these children with the chilling caption "Youths who put God first."

These completely avoidable deaths continue to mount. In 1998, Seth Asser and Rita Swan tried to measure the toll in a paper published in the medical journal *Pediatrics*. Their aim was to determine how many children had died from religiously based medical neglect in the twenty years after 1975, and how many could have been saved. To that total they added mortalities of fetuses and infants during and shortly after birth when doctors and midwives were barred on religious grounds. Of course, determining after the fact whether medical intervention would have saved lives is a judgment call, but in many cases, including childhood diabetes, ruptured appendixes, and breech births, medical intervention is nearly always successful.

The results were both startling and depressing. Of 172 children who died over those two decades after being denied medical care on religious grounds, 140—81 percent of the total—had conditions that would have been curable with a probability of greater than 90 percent. Another 18 (10 percent) had a probability of cure greater than 50 percent but less than 90 percent. Only 3 (victims of a car accident, a severe heart defect, and anencephaly) would not have benefited from medical attention. Asser and Swan's list of examples is heartbreaking; here are but three:

> For example, a 2-year-old child aspirated a bite of banana. Her parents frantically called other members of her religious circle for prayer during nearly an hour in which some signs of life were still present.

> One teenager asked teachers for help getting medical care for fainting spells, which she had been refused at home. She ran away from home, but law enforcement returned her to the custody of her father. She died 3 days later from a ruptured appendix.

> One father had a medical degree and had completed a year of residency before joining a church opposed to medical care. After 4 days of fever, his 5-month-old son began having apneic episodes. The father told the coroner that with each spell he "rebuked the spirit of death" and the infant "perked right back up and started breathing." The infant died the next day from bacterial meningitis.

Both mothers and fetuses have died after refusing to have doctors or midwives present at childbirth. Is there anything other than faith—or complete ignorance of medicine—that could have caused the following gruesome scene?

In one case, a 23-year-old woman presented to an emergency room after 56 hours of active labor with the infant's head at the vaginal opening for 16 hours. The dead fetus was delivered via emergency cesarean, and was in an advanced state of decomposition. The mother died within hours after delivery from sepsis because of the retained uterine contents. The medical examiner noted that the corpse of the infant was so foul smelling that it was inconceivable anyone attending the delivery could not have noticed.

The believers in these cases were not just Christian Scientists (16 percent of the total deaths), but represented twenty-three Christian denominations from thirty-four states.

Such deaths are unconscionable because they involve children who have no say—or no mature say—in their own medical care, but are at the mercy of their parents' beliefs. Because injuring a child by withholding medical care for *nonreligious* reasons constitutes legal child abuse, it's hard to make the case that it's not equally abusive when medical care is rejected on religious grounds. In such light, Jesus's statement in Matthew (19:14)—"Suffer little children, and forbid them not, to come unto me: for of such is the kingdom of heaven"—has a horrible double meaning.

It's not just the parents who are at fault. Religious exemptions are written into law by the federal and state governments—that is, those who represent all Americans. In fact, thirty-eight of the fifty states have religious exemptions for child abuse and neglect in their civil codes, fifteen states have such exemptions for misdemeanors, seventeen for felony crimes against children, and five (Idaho, Iowa, Ohio, West Virginia, and Arkansas) have exemptions for manslaughter, murder, or capital murder. Altogether, forty-three of the fifty states confer some type of civil or criminal immunity on parents who injure their children by withholding medical care on religious grounds.

Surprisingly, these exemptions were *required* by the U.S. government in 1974 as a condition for states to receive federal aid for child protection. Before that, only eleven states had such exemptions; afterward there were forty-four. (That requirement was rescinded in 1983, but it was too late: most states had enacted the religious exemptions, which are still in place.)

The government, or rather the taxpayers, further support religious child abuse by subsidizing Christian Science practitioners and their nursing homes with Medicare and tax exemptions—despite their complete failure to provide *any* medical care. Other tax support involves allowing federal employees, some state employees, and members of the armed forces to join health plans that include Christian Science nursing and practitioner care.

The tangle of laws, in which parents in some states can be exculpated from abuse or neglect but convicted of manslaughter, has led to mass confusion in the courts. The result is that when parents are found guilty of medically neglecting their children on religious grounds, their convictions can be thrown out of court because of conflicting laws. Or when parents *are* convicted, religiously based sympathy for them results in trivial punishments, usually probation or a small fine. Only rarely do such parents go to jail. This unwarranted sympathy for faith eliminates the stiff sentences that might deter other parents from using religious healing on their children.

It's not just medical care that's subject to religious exemption in America. You can, on religious grounds alone, refuse to have your newborn children tested for metabolic diseases or given prophylactic eyedrops. You can refuse to have your children tested for lead in their blood. In Oregon and Pennsylvania there are even religious exemptions from wearing bicycle helmets. In California, public school teachers can refuse to be tested for tuberculosis on religious grounds, which, of course, could endanger their students.

Religious exemptions for vaccinations, allowed in forty-eight of the fifty U.S. states (all except Mississippi and West Virginia), endanger not only the children who don't get immunized, but the community in general, for even those who are vaccinated don't always acquire immunity. To attend public schools and many colleges, like the one where I teach, students must show evidence of vaccination for diseases like hepatitis, measles, mumps, diphtheria, and tetanus. The only exemptions permitted are for medical reasons, like a compromised immune system—and religion.

Nor are Christians the only believers who oppose immunization. Islamic clerics in Afghanistan, Pakistan, and Nigeria urge their followers to oppose polio vaccination, declaring it a conspiracy to sterilize Muslims. These efforts may prevent the complete eradication of polio from the human species, something already achieved for smallpox. Dr. A. Majid Katme,

spokesman and former head of the Islamic Medical Association of the
United Kingdom, described by the *Guardian* as "a respected figure in the
British Muslim community," has come out against *all* childhood vaccination, claiming that "the case of vaccination is first an Islamic one, based on
Islamic ethos regarding the perfection of the natural human body's immune
defense system, empowered by great and prophetic guidance to avoid most
infections." Taking his advice would, of course, be disastrous.

Few records are kept on *adult* deaths caused by religious "healing," but
we can get an idea of the problem from a study of childbirth in women belonging to the Faith Assembly, an antimedical religious group in Indiana.
Not only was mortality of late fetuses and early newborns three times
higher than for the state as a whole, but the *maternal* mortality rate during
childbirth was ninety-two times higher!

Of course, nobody considers prosecuting adults who favor spiritual
treatment for themselves over modern medicine, as they are presumed capable of making their own decisions, however foolish. Yet their choices may
not be as free as we think. For many of the parents who withhold medical
care from their children once *were* those children, raised in the faith and
indoctrinated in its tenets. Of course, the Kings should have been punished
more severely than they were, both to prevent them from repeating their
behavior and to deter others from imitating it, but it's hard to argue that
parents raised in such a faith are completely free to reject what was drilled
into them when they were young and credulous.

I have dwelt on medical exemptions because they clearly show the conflict between science and faith, as well as the grievous harm that this can
cause. Medicine can cure; faith cannot. But such faith need not be religious.
In 2013, Tamara Sophie Lovett, a Canadian, was charged with negligence for
treating her seven-year-old son Ryan, severely infected with streptococcus,
with homeopathic and herbal remedies. Although such infections can usually be cured easily with a dose of penicillin, Ryan didn't have that option,
and died. What killed him was not religious faith, but faith in alternative
medicine. As one investigating officer said, "We have no direct information
that religious beliefs factored into this, but there was a belief system and homeopathic medicine did factor in." Faith is faith, and in this case it too conflicted with science.

While the conflict between creationism and evolution reduces Americans' scientific literacy, nobody dies from not learning about evolution. Faith-based healing is a different matter. A single child killed in the name of faith is one too many. How many will it take before we realize that our exaggerated respect for religion will allow these deaths to continue indefinitely? The case of the Kings, and others like it, is a good example of Steven Weinberg's argument that "for good people to do evil—that takes religion"—or sometimes just faith.

Many of the parents who injure or kill their children through medical neglect are not toothless Bible-thumpers, nor even biblical fundamentalists. Many, like Christian Scientists, can even be seen as religious "moderates"—a group that, accommodationists claim, are relatively harmless, or even helpful as allies in the fight against creationism.

But Americans of even liberal faiths—which can include the legislators who write religious exemptions and tax breaks into laws, the believers who lobby for such laws (even if they don't endorse faith healing), and the prosecutors and judges who are reluctant to prosecute such cases, or dole out light sentences upon conviction—all are complicit in the death of those children. Sadly, the *real* guilty parties, the churches that promulgate faith healing, are always exempt from punishment. Political pressure—the "anti-religion" characterization that would attach to American politicians who lobby against religious exemptions—prevents any movement to repeal these unfair and harmful laws.

Why does this continue? Faith enables other faith, of course, and there's "belief in belief," the idea that religions, regardless of their tenets, should be encouraged as a social good. Religious lobbies such as the Christian Science church, which has always pushed hard for medical-care exemption laws, strengthen this attitude. Further, many Americans already use prayer as an adjunct to regular medicine. Within a given year, for instance, between 35 percent and 62 percent of adults employ prayer for health reasons. If you think that prayer can *supplement* regular medicine, it becomes easier to accept the view of some sects that prayer can *substitute* for regular medicine. And some "mainstream" faiths hold views that come perilously close to faith healing. The Vatican, for instance, has an official exorcist—Gabriele Amorth, who claims to have performed the rite over seventy thousand times—and the Catholic Church recently gave formal recognition to the International Association of Exorcists, comprising 250 priests in thirty countries. How many Catholics are aware that their church officially recognizes demonic possession and has procedures to deal with it? And who knows how many disturbed people have been subjected to a frightening procedure that is harmless at best, but potentially dangerous, especially when those who use it misconstrue and thus ignore the real causes of mental illness?

Even scientists have given their imprimatur to faith-based healing. In 1992, the U.S. Congress funded an Office of Alternative Medicine, which seven years later became the National Center for Complementary and Alternative Medicine (NCCAM), still associated with the prestigious National Institutes of Health. In the two decades ending in 2012, the government sank $2 billion into NCCAM. Despite that huge expenditure, the center has never produced one bit of evidence for the value of "alternative medicine"— and that includes acupuncture, reiki, and various forms of spiritual healing. (The joke among advocates of scientific medicine is "What do you call alternative medicine that works? Medicine.") The work funded by NCCAM included studies on the effects of "distance healing"—including prayer—on HIV and glioblastoma (brain cancer), on coffee enemas as a palliative for cancer, and on magnetic mattress pads as cures for arthritis. None of these studies gave positive results; indeed, many of their results haven't even been published.

While "alternative medicine" is often secular, it bears many similarities to religion, for they share many characteristics of pseudoscience. Practitioners of both tend to ignore counterevidence or reject it with special pleading; embrace unfalsifiable claims ("It won't show up in double-blind tests," equivalent to "You can't test God"); accept questionable data as "proof"; argue that the scientific method doesn't apply to their claims; reject replication and verification by outsiders and skeptics; and refuse to consider alternative hypotheses. Above all, religion, faith healing, and alternative medicine all show the diagnostic feature of faith: an agenda not to find the truth, but to support one's biases, emotions, and personal beliefs. In fact, the growing popularity of "integrative medicine"—a mélange of traditional medicine and more "holistic" therapies—can be seen as a form of accommodation between science and spirituality.

In the preceding chapter I argued that, unlike science, religion comes with a built-in plan of action, for if your beliefs are wedded to both a moral code and the promise of eternal reward or punishment, you may feel the duty to inculcate not only your children with those beliefs, but others as well. Rodney Stark, a Christian sociologist of religion, compares the duty to spread one's beliefs with the duty to disseminate a lifesaving vaccine:

> Imagine a society's discovering a vaccine against a deadly disease that has been ravaging its people and continues to ravage people in neighboring societies, where the cause of the disease is incorrectly attributed to improper diet. What would be the judgment on such a society if it withheld its vaccine on the grounds that it would be ethnocentric to try to instruct members of another culture that their medical ideas are incorrect, and to induce them to adopt the effective treatment? If one accepts that one has the good fortune to be in possession of the true religion and thereby has access to the most valuable possible rewards, is one not similarly obligated to spread this blessing to those less fortunate? . . . Only One True God can generate great undertakings out of primarily religious motivations, and chief among these is the desire, indeed the duty, to spread knowledge of the One True God: *the duty to missionize is inherent in dualistic monotheism.*

Suppression of Research and Vaccination

Religious missionizing doesn't always involve visiting people's houses to let them in on the Good News. It usually takes the more subtle form of trying to force others to conform to your moral beliefs through political lobbying or trying to make new law. When such political action rests on religious beliefs that conflict with science, as it has with medical-care exemption laws, that means that both science and society suffer. This has happened, and is still happening, in several other areas where science clashes with faith in the public arena.

One involves research on embryonic stem cells (ESSs), cells that are initially undifferentiated but have the capacity to develop into a variety of different tissues. That gives them enormous potential for curing human diseases, regrowing tissues and organs, and attacking a variety of previously intractable medical problems like Alzheimer's disease, Parkinson's disease, and spinal cord injuries. The cells are usually extracted from early-stage embryos (less than a week old) that have been frozen as a surplus product of in vitro fertilization, which involves producing many embryos in the laboratory. When thawed, those embryos can be implanted in a woman's uterus, usually with about a 50 percent success rate. Extra embryos are made in case the first implantation is unsuccessful.

There are hundreds of thousands of these frozen embryos languishing in cylinders of liquid nitrogen, never to be implanted in women. They are simply leftovers. But although their cells have enormous potential to help living humans, both children and adults, their use has been restricted for years—because of faith. Many religions hold that a single fertilized egg is equivalent to a person, and that destroying early-stage embryos is therefore murder. It is because of this opposition that in 1995, President Bill Clinton, acting against the advice of a panel from the National Institutes of Health, banned federal funding for any research involving the destruction of frozen human embryos. Six years later, President George W. Bush restricted the expansion of the few ESS lines approved for research, limiting the number to just twenty-one. In 2006, Bush used his first presidential veto to override a congressional bill allowing federal funding of embryonic stem cell research. A born-again Christian, Bush gave clearly religious reasons, characterizing the bill as one that "would support the taking of innocent human life in the hope of finding medical benefits for others," adding that "it crosses a moral boundary that our decent society needs to respect."

The situation has improved a bit since President Obama expanded federal funding for ESS lines and increased the number of lines supported by that funding from twenty-one to sixty-four. The use of "pluripotent" stem cells from adults, as well as private funding, has helped fill the gap. But the number of lines is still limited—clearly on moral grounds—and some states like Louisiana and the Dakotas still ban *all* ESS research. There is little doubt that without religiously based opposition, stem cell research would be several years ahead of where it is. And that almost certainly means that some humans will die needlessly.

One of the most egregious forms of religious opposition to science involves vaccination against the human papillomavirus (HPV), the major cause of cervical cancer in women, but also a cause of cancers of the anus, vulva, vagina, and pharynx, as well as genital warts. Spread through sexual contact, the infection can be almost completely prevented with a series of three safe injections, and is recommended by the Centers for Disease Control and Prevention for boys and girls eleven or twelve years old. Cervical cancer is deadly: it causes about four thousand deaths per year in the United States and almost two hundred thousand worldwide, most of which could be prevented by vaccination of preteens.

In the United States, both religious and right-wing political organizations (their membership shows considerable overlap) often favor the availability of anti-HPV vaccines, but strongly oppose attempts to make immunization mandatory. (Surprisingly, in 2007 Governor Rick Perry of Texas, a religious and conservative Republican, signed an executive order making vaccination mandatory for young girls in his state, but it was overturned by the state legislature. Perry, probably under pressure from believers, later admitted, "If I had it to do over again, I would have done it differently.")

Although HPV is spread through sexual contact alone, given the dire consequences of cervical cancer, and the vaccination's known safety and effectiveness, there is no reason not to make it as mandatory as vaccinations for measles and diphtheria. That is, there is no reason save one: by eliminating one dangerous side effect of sex, the vaccination supposedly encourages promiscuity, both pre- and postmarital. And many Christians oppose any sex outside of marriage. Despite studies showing no apparent increase in sexual activity following HPV vaccination, religious people still oppose requiring it. In a position paper on the vaccine, for instance, Focus on the Family, a conservative American religious organization, favored abstinence over vaccination:

> The seriousness of HPV and other STIs [sexually transmitted infections] underscores the significance of God's design for sexuality to human well-being. Thus Focus on the Family affirms—above any available health

intervention—abstinence until marriage as the best and primary practice in preventing HPV and other STIs.

As one might expect given its view of nonmarital sex as sinful, the Catholic Church has been one of the biggest opponents of mandatory HPV vaccination, and not just in the United States. Although the Canadian government provides free vaccinations for children, the Catholic bishop of Calgary, Fred Henry, publicly opposed endorsement of the vaccine by Catholic school boards in Canada—schools funded by the government—because it would "undermine the schools' effort to teach children about abstinence and chastity in accord with the teachings of the Catholic Church." Pressure from the Catholic Church has also halted the administration of anti-HPV vaccines in Trinidad and Tobago, with the church going so far as to question, in the face of all scientific fact, the safety of the vaccine.

It's well documented that abstinence vows and abstinence-based education are largely useless at reducing teen pregnancy and the incidence of sexually transmitted diseases. Yet even if such programs were partly effective, those for whom they don't work are still at risk for infection. The cost of three injections is much less than that of treatment for HPV-induced cancer. Given the sexual activity of young people, and the risk involved in sex with an infected person, the prevention of HPV is a more severe public health problem than that of measles, for which—except for some religions—vaccination is required. Given the exaggerated respect for faith in the United States, it's certain that even were HPV vaccination required for schoolchildren, there would still be exemptions based not on any dangers of the vaccine but, given its supposed (but false) encouragement of sex, on religious belief. Parents who refuse to vaccinate their sons and daughters for HPV are making a conscious decision to let their children risk death if they have premarital sex. They are kin to the Christian Scientists, Jehovah's Witnesses, and faith healers who martyr their children to their faith, and to the Catholic hospitals that allow both a mother and her baby to die in childbirth rather than offer an abortion.

Religion further impedes treatment of disease by imputing epidemics to God, implying that the cures lie in correcting immoral behavior rather than in medical attention. (This is similar to Christian Science on a large scale.)

In America we regularly hear that earthquakes, tornadoes, and droughts are due to God's wrath about, say, homosexual behavior or abortion. As I write, the Ebola virus is ravaging West Africa, and there's no telling whether it will be stemmed or will spread widely. But the situation isn't helped by local religious leaders, including the Liberian Council of Churches, which blamed corruption and immorality, including "homosexualism," for the epidemic. This led Liberia's president, Ellen Johnson Sirleaf, to call for three days of prayer and fasting to ask for God's mercy. This has only fueled Liberians' suspicion about Western medicine, leading them to refuse medical attention and turn away government workers. We're seeing a return to medieval days when the Black Death that ravaged Europe was seen as God's punishment for sin. The difference is that now we have medical treatment to help cure Ebola, and epidemiology to plot the best strategy to attack it.

Opposition to Assisted Dying

What other harms spring from the morality claims of faith, claims that flout both science and reason? One is opposition to assisted dying. It's hard to believe that a world without religion would have any problems with a regulated system to help the terminally ill end their lives. After all, most of us consider it merciful to euthanize our dog or cat if it's suffering terribly with no hope of respite. Nobody would consider it a moral act to let such animals suffer because only God has the right to end their lives. And yet this is precisely how many religions behave toward humans, for humans are exceptional—the special creation of God, and uniquely endowed with souls.

If you've had your animal "put to sleep," then you know the process is humane and painless. And now science has ways to allow humans to end their lives painlessly as well: an overdose of pentobarbital works effectively, and is used in European countries, like Switzerland, that permit assisted dying. Who would prohibit a terminal patient, suffering from cancer or a neurodegenerative disease, from deciding to take that route rather than suffer needlessly for months?

Many religious denominations would. Although secular society is gradually recognizing that assisted dying in terminal cases is not only merciful but moral, some faiths, especially Catholicism, object on the grounds that only God has the right to determine when you will die. (As with most issues this controversial, many liberal Catholics disagree with church policy.) In its official Declaration on Euthanasia, the church affirmed that assisted suicide is actually sinful in two ways. First, the decedent himself sins by suicide, a sin that may, depending on God's mood, send one to hell:

> Intentionally causing one's own death, or suicide, is therefore equally as wrong as murder; such an action on the part of a person is to be considered as a rejection of God's sovereignty and loving plan.

Although it's seen by nearly everyone as humane—and even moral—to end the life of our terminally ill pets, it's regarded as murder to make the same decision for ourselves. This unconscionable equation of murder with end-of-life assistance clearly has a religious basis, for the church sees unbearable suffering as an actual good:

> According to Christian teaching, however, suffering, especially suffering during the last moments of life, has a special place in God's saving plan; it is in fact a sharing in Christ's passion and a union with the redeeming sacrifice which He offered in obedience to the Father's will.

Is there any institution other than religion that could see terminal suffering as beneficial?

Further, the church declares that anyone who, without repenting, hastens death for the terminally ill is also bound for damnation. In a 1995 church document, Pope John Paul II declared:

> I confirm that euthanasia is a grave violation of the law of God, since it is the deliberate and morally unacceptable killing of a human person. This doctrine is based upon the natural law and upon the written word of God, is transmitted by the Church's Tradition and taught by the ordinary and universal Magisterium. . . . Depending on the circumstances, this practice involves the malice proper to suicide or murder.

("Natural law," of course, does not refer to the scientific "laws of nature," but to the innate morality supposedly vouchsafed to Catholics by God and understood by reason.)

With its cult of suffering, the Catholic Church has clearly taken the lead on this issue, but other faiths have followed. Anglicans in Canada and the United Kingdom are opposed to assisted dying, as are Baptists and other evangelical Christians in America. Protestants, Buddhists, and Jews are divided, while Muslims, based on Muhammad's supposed opposition to suicide, are adamantly opposed to assisted dying, holding both the decedent and those who aid his death culpable. In most countries these religious views are the law, and those who lack money to travel to the few countries that allow assisted dying are doomed to perish in agony. (In the United States, only five states—Oregon, Washington, Montana, New Mexico, and Vermont—permit physicians to help patients end their lives.) While many nonreligious people would still resist assisted dying, organized religion has thrown its weight behind trying to deny people that choice.

Global-Warming Denialism

The ability of people to ignore inconvenient truths that conflict with their faith, whether or not the faith be religious, is astonishing. An AP-GfK poll taken in 2014 showed that fully 51 percent of Americans were either "not too confident" or "not at all confident" in the universe's having begun 13.8 billion years ago with the Big Bang, something that all cosmologists accept. Forty-two percent showed a similar lack of confidence that life on Earth (including humans) evolved by natural selection, 36 percent lacked confidence in the indisputable fact that the Earth is 4.5 billion years old, and 37 percent weren't confident that the Earth's temperature was rising because of man-made greenhouse gases. When people's religious beliefs were tallied with those data, the pollsters got the unsurprising result that acceptance of evolution, the Big Bang, the Earth's age, and anthropogenic global warming was dramatically lower among those who were more confident about God's existence and who attended church more often. Given this correlation with religious faith, and the widespread availability of evidence for the age of the universe and Earth, evolution, and anthropogenic global warming, it's hard to believe that all of this ignorance reflects a lack of acquaintance with scientific facts.

While such ignorance is distressing, it becomes positively harmful when coupled with political and social action. Such is the case with global-warming denialism. But while doubts about evolution and the age of the universe usually rest on conflicts between scientific and religious views of nature, it's not so clear what climate-change denialism has to do with religion. As we'll see, there is a connection, one based on beliefs about God's stewardship of the planet and his promise to preserve it until his return.

While denial of evolution doesn't pose an immediate danger to the planet, denial of global warming does. It is now the nearly unanimous view of climate scientists that the Earth is warming because of human-generated emissions of greenhouse gases like carbon dioxide and methane. Unless these emissions are curtailed, the results will be dire not just for humans (coastal areas will be inundated, droughts will reduce food supplies), but for other species whose habitats will change or vanish. We will, for example, certainly see the demise of species like the polar bear.

Nothing less than the future of our planet is at stake, and denialist opposition is simply dangerous. That opposition comes in two forms, both involving pseudoscience. One rests on manufactured "scientific" claims supporting preconceived political positions, the other on religious belief. Both in the end rely on faith, and on the need to hold views that agree with those of your group, be it religious or political.

And just as we saw for the religiously motivated opponents of evolution who were just as science-savvy as those who accept evolution, climate-change denialists also know just as much about the relevant science as those who accept global warming. Yet denialists dismiss that science because it conflicts with the beliefs of their "community"—usually political conservatives. As Dan Kahan, a professor of law and psychology at Yale University, found when analyzing earlier data:

> When people are shown evidence relating to what scientists believe about
> a culturally disputed policy-relevant fact (e.g., is the earth heating up? is

it safe to store nuclear wastes deep underground? does allowing people to carry hand guns in public increase the risk of crime—or *decrease* it?), they selectively credit or dismiss that evidence depending on whether it is consistent with or inconsistent with their cultural group's position. As a result, they form polarized perceptions of scientific consensus even when they rely on the same sources of evidence.

These studies imply misinformation is not a decisive source of public controversy over climate change. People in these studies are *misinforming themselves* by opportunistically adjusting the weight they give to evidence based on what they are already committed to believing.

The only difference between this case and that of evolution is that the communities are political in the former and religious in the latter.

Religion's role involves declaring that God gave us stewardship over the Earth, which some interpret as also allowing us to do whatever we want with the planet. This is often coupled with the claim that God will ensure that the Earth will recover from all human intrusions. So, for example, Rick Santorum, a Republican politician and former U.S. senator, proclaimed at a meeting in Colorado that climate change was a "hoax," and then played the religion card:

We were put on this Earth as creatures of God to have dominion over the Earth, to use it wisely and steward it wisely, but for our benefit not for the Earth's benefit. . . . We are the intelligent beings that know how to manage things and through the course of science and discovery if we can be better stewards of this environment, then we should not let the vagaries of nature destroy what we have helped create.

John Shimkus, a congressman from Illinois, went even further, quoting from the Book of Genesis when testifying in 2009 before the House Subcommittee on Energy and Environment:

"Never again will I curse the ground because of man, even though every inclination of his heart is evil from childhood, and never again will I destroy all living creatures as I have done. As long as the Earth endures, seed time and harvest, cold and heat, summer and winter, day and night will never cease" [Genesis 8:21–22]. I believe that's the infallible word of God, and that's the way it's going to be toward his creation. . . . The Earth will end only when God declares it's time to be over. Man will not destroy this Earth. This Earth will not be destroyed by a flood. I appreciate having panelists here who are men of faith, and we can get into the theological discourse of that position. But I do believe God's word is infallible, unchanging, perfect.

It's unbelievable that an elected official can try to affect government policy by twisting a quotation from an Iron Age deity. And Shimkus is not alone. Thirty-six percent of his fellow Americans (and 65 percent of white evangelical Protestants) see increasingly severe natural disasters as evidence for the arrival of the biblical End Times, which, of course, are part of God's plan. Slightly more Americans (38 percent) agree that "God gave human beings the right to use animals, plants, and all the resources of the planet for human benefit."

The most blatant denial that the Earth is endangered comes from the "Evangelical Declaration on Global Warming" by the Cornwall Alliance for the Stewardship of Creation, a conservative Christian think tank. Here's part of it:

What We Believe
We believe Earth and its ecosystems—created by God's intelligent design and infinite power and sustained by His faithful providence—are robust, resilient, self-regulating, and self-correcting, admirably suited for human flourishing, and displaying His glory. Earth's climate system is no exception. Recent global warming is one of many natural cycles of warming and cooling in geologic history. . . .

What We Deny
We deny that Earth and its ecosystems are the fragile and unstable products of chance, and particularly that Earth's climate system is vulnerable to dangerous alteration because of minuscule changes in atmospheric chemistry. Recent warming was neither abnormally large nor abnormally

rapid. There is no convincing scientific evidence that human contribution to greenhouse gases is causing dangerous global warming.

That statement was signed by hundreds of ministers, theologians, physicians, scientists, economists, and educators, even including prominent meteorologists like Joseph D'Aleo and Neil Frank, former director of the National Hurricane Center. The document calls for continuing use of fossil fuels and makes dire predictions of how "the poor" will be harmed by reducing harmful emissions.

To show how far religion (or simple wish-thinking) can affect scientific judgment, another part of the declaration makes the completely insupportable statement that "we deny that carbon dioxide—essential to all plant growth—is a pollutant." Yes, carbon dioxide is consumed by plants during photosynthesis, but it's also produced in huge quantities by burning fossil fuels. Water is also essential to plant growth, but forests can be destroyed by flooding, and cities by rising sea levels. Carbon dioxide is in fact the main greenhouse gas that causes global warming (others are methane and nitrous oxide), and denying that is not only willful ignorance, but a recipe for disaster.

While one can claim that the religion card is played only as window dressing for the true motivation of greediness (or suspicion of scientists), there's no reason to think these believers aren't sincere. The common religious view that God will save the Earth, or that pollution doesn't matter because the End Times are at hand, surely affects how people regard global warming.

Because Americans with conservative attitudes tend to be religious (particularly when they're more involved in politics), it's often difficult to separate views on climate based on religion from those resting on a secular faith-based rejection of science. This shows again how pseudoscience converges with religion. And even when not motivated by religion, climate-change denialists still make palpably false claims resembling those used by advocates of alien abduction or Holocaust denialism. Climate denialists have, for example, asserted that scientists on a climate-change panel of the U.S. National Academy of Sciences, whose report implicated fossil fuels in global warming, actually profited financially from their efforts (not true: they don't get a penny for such work). Other arguments are that climate-change scientists don't base their conclusions on "real scientific facts"; that the "real" evidence shows no trend of global warming, which is "one of the greatest hoaxes perpetrated out of the scientific community . . . there is no scientific consensus"; and that climate-change concern is "a massive international scientific fraud." Amazingly, all of these quotations come from Republican members of the House of Representatives Committee on Science, Space, and Technology, the committee responsible for formulating U.S. policy on such issues. Fully 72 percent of the committee members are outright climate-change denialists or have voted against bills to alleviate global warming.

In the end, these attitudes, and their political consequences, come from believing what you'd like to be true rather than what science tells us. Nowhere have I seen this more clearly expressed than by a man who should know better: Brother Guy Consolmagno, a Jesuit astronomer and curator of meteorites at the Vatican Observatory. Noting people's surprise that the Vatican even *has* an observatory, Consolmagno explained that it serves "to make people realize that the church not only supports science, literally . . . but we support and embrace and promote the use of both our hearts and our brains to come to know how the universe works." It's hardly necessary to add that the heart is not an organ for thinking, and that we can never understand "how the universe works" by using our emotions, via faith, instead of reason. Father Consolmagno's heart, for example, has convinced him that if extraterrestrial beings exist, they have souls like ours.

Does Faith Have Any Value?

It's intriguing to contemplate how the world would be different without empirical beliefs based on faith—and not just religious faith. What would that be like?

What we would lose in a world without faith would not be the good things—the art and literature, the fellow feeling that inspires us to help others, the moral impulses (as we'll see, Europe is largely nontheistic but hardly a hotbed of immorality)—but the bad ones. On the secular side, we wouldn't have homeopathy or other nonreligious forms of "alternative medicine," and there would be less opposition to global warming and vaccination. Debates about abortion, universal health care, and much of politics would be far more informed by facts, though, of course, they'd still involve subjective preferences.

On the religious side, we'd lose the harmful tenets of belief that rest on the certainty of God-given morality: the dysfunctional aspects of society that in the absence of religion would find little support. Catholicism is seen as one of the less extreme faiths, yet if its beliefs didn't rest on scripture, this is what would diminish: opposition to abortion, euthanasia, and stem cell research; opposition to divorce; belief in the sinfulness of homosexuality; cramping of the sex lives of consenting adults; the second-class status of women (at least in the church) and the notion that they're largely vehicles for producing more Catholics; opposition to birth control and HPV vaccination; the incidence of AIDS; and the terrorizing of children with guilt and threats of eternal damnation.

And that's just one faith. Islamic tenets are even more harmful because the tight connection between Islam and politics means that beliefs are directly converted into law—often sharia law. Sharia comes straight from the Quran, as well as the hadith and the sunnah—the reputed sayings, teachings, and beliefs of Muhammad. The law thus takes its authority from Muhammad's status as a conduit for Allah, an empirical claim that for most Muslims is simply beyond dispute.

While the interpretation of sharia law and the degree to which it's applied in civil versus criminal matters vary among Islamic states, some, like Saudi Arabia, Iran, Yemen, and Sudan, embrace it completely, becoming theocracies. Many other countries—including Egypt, Libya, Algeria, India, Pakistan, and, surprisingly, Israel—allow it only to Muslim residents, and only for special matters like marriage and inheritance. Here is what sharia law mandates: the subjugation of women (including the legality of child brides), as well as inheritance and divorce laws that discriminate against women (in a sharia court, a woman's testimony counts only half as much as a man's, and a conviction for rape requires a woman to produce *four* witnesses of it). Sharia law dictates corporal punishment like lashing or amputation for theft, and prescribes the death penalty for both homosexuality and apostasy. Every one of the thirteen countries that impose capital punishment for rejecting the state religion or espousing atheism is Islamic. Can there be any doubt that had Islam never existed, much of this irrationality would not be with us? (How, for instance, can you kill someone for apostasy if there's no faith to leave?) And so it is with numerous other faiths, many of which use scripture to sanction misogyny and hatred of minorities, as well as to regulate the diets, clothing, and sex lives of their adherents.

Finally, we'd lose a lot of the divisiveness that threatens to tear our world apart. Muslim against Christian and Jew, Hindu against Muslim, Buddhist against Hindu, Catholic against Protestant, Sunni against Shiite—all hatred based solely on faith would disappear. Of course, there would still be strife and xenophobia, which probably rest largely on evolution, but can you really claim that hatred based on religion would inevitably be replaced by hatred based on something else, as if the world had to fulfill a given quota of enmity? After all, Sunnis and Shiites are still Muslims, and have the same cultural background. They kill each other for faith alone.

And there is no reason that a world without faith, particularly religious faith, would be a programmed Stalinesque society, like a hive of bees. We know this because, as we'll see shortly, the largely nonreligious societies of Europe are good ones, certainly more livable and perhaps more moral than those—including Western religious cults like the Amish—that are essentially theocracies.

I've argued that pure faith of any stripe, be it in God, homeopathy, or ESP, should be rejected. But is that always the case? Could it ever be good to have faith? That is, are there important situations—ones on which your life or well-being heavily depend—when you should act in the absence of information you *could* have obtained, or *against* relevant information? Are there situations in which you should guide yourself by wish-thinking, revelation, or unevidenced dogma?

My answer is "sometimes, but not very often"—and there are a few caveats that go with that answer. By "faith" I mean, as always, belief without verifiable evidence. And, of course, the answer to the question of "is it good?" is not simple, for one must distinguish what's good for the believer from what's good for others, or for society as a whole.

A common example of a supposedly useful faith is the "dying grandmother" scenario. Your grandmother is on her deathbed, and is deeply consoled by thinking that she'll soon be in heaven, reunited with her late husband and ancestors. You don't believe a bit of it, but refrain from saying anything. What's wrong with that?

Nothing. If that faith eases her last moments, it would be churlish to attack it, for the costs are high and the benefits nil. Unfortunately, this scenario is often used as a criticism of atheists, who, say critics, are supposedly champing at the bit to dispel the poor woman's illusions about an afterlife. But I know of no nonbeliever who would sanction that, or say there's anything wrong with allowing the dying to retain their faith. In fact, it is *theists* who try to convert people on their deathbeds, informing the terminally ill that they'll burn in hell unless they accept Jesus. As a prominent atheist, the late Christopher Hitchens was particularly subject to this form of harassment.

But while having that kind of faith might be beneficial at the end of a believer's life, that doesn't mean that *society in general* is better for having such faith. As we've seen, there are strong arguments against that. Apart from the harmful effects of buying into a religious morality, most people live a long time before they die, and for many their lives would differ substantially if they didn't believe they were facing eternal reward or punishment.

As for having faith, religious or otherwise, that you'll beat a life-threatening illness, there's little harm in that—with one exception. While such optimism may stave off depression (though it doesn't seem to help much with cures), it has the bad side effect of putting off your preparations for the likely result: making peace with old enemies, saying good-bye to loved ones, putting one's affairs in order, and so on. Although religion can buy consolation, it often does so at the expense of practicality.

This trade-off between consolation and practicality is important when considering one of the most common arguments for the value of faith—one used by both atheists and believers alike. Even if we have little or no evidence for the divine, the argument goes, it's still beneficial for people to believe in a god. This argument takes two forms, depending on the population you're considering. If you're thinking of society as a whole, one can argue that faith is a critical social glue, bonding us together in solidarity, comity, and morality. If you're considering only the downtrodden, marginalized, or poor, you can claim that religious belief gives such people hope and a reason to go on—often because they believe all will be set right in the afterlife. And even if that hope is specious and death is final, they'll never know the difference, but their life will have been less torturous.

Both of these notions exemplify what Daniel Dennett calls "belief in belief": the claim that faith doesn't necessarily have to be true to be useful. I've heard many fellow atheists make this argument, which from their mouths sounds deeply condescending: "We're sophisticated enough to dispense with gods, but the Little People must have theirs. After all, they're not susceptible to reasoned argument, and can't be fulfilled without faith." But even coming from the faithful, the first argument, that religion is a social necessity and will always be with us, is dubious. It can be demolished with only two words: northern Europe.

Once deeply religious (Spinoza, after all, was expelled from Amsterdam's Jewish community for heresy), northern Europe has in the last few centuries become largely atheistic. The degree of pure atheism—those who agree with the statement "I don't believe there is any sort of spirit, god, or life force"—runs between 25 and 40 percent of the population of countries like Germany, Denmark, France, and Sweden. The level of nontheistic spirituality ("I believe there is some sort of spirit or life force, but not God") is even higher: 25 to 47 percent. When you add the two groups to get the total proportion of nontheists, it's a majority: 71 percent in Denmark, 73 percent in Norway, 79 percent in Sweden, 67 percent in France, 52 percent in Germany, and 58 percent in the United Kingdom. In contrast, nontheists in America constitute only 18 percent of the population, while 80 percent affirm a belief in God.

Now, there's no evidence that northern Europe is socially dysfunctional. In fact, one could make a good case that in many ways those nations function better than does the highly religious United States. Sociologists measure the well-being of countries using indices of social dysfunction that include things like levels of divorce, homicide, incarceration, juvenile mortality, alcohol consumption, poverty, income inequality, and so on. And on those scales Scandinavia and northern Europe rank much higher in well-being than the United States, which among seventeen First World countries surveyed was dead last. (The four most "successful" societies were Norway, Denmark, Sweden, and the Netherlands.) My own analysis further showed a negative relationship between societal well-being and religiosity: the least religious societies were the most successful. While this correlation doesn't by itself implicate a cause, neither does it support the claim that religion is essential for a harmonious society. Nor does northern Europe appear to be a hotbed of immorality, despite the claim that religion both furnishes and enforces morality.

Although Europeans' reliance on faith must surely have waned over the last few centuries, their social harmony doesn't seem to have suffered. While Europe's experience cannot necessarily be taken as universally applicable, and secularization has been much slower in the United States, the data fail to support the claim that faith is both inevitable and necessary for a well-ordered society.

The second argument for faith is that it gives solace to the marginalized and destitute. And that's no doubt true. When you see yourself as being without hope, there is consolation in thinking that God and Jesus are looking out for you (even if they're not helping much), and in thinking that all will be set right in the next world. I suspect that's why European countries with strong social safety nets—including government-sponsored medical care, paternity and maternity leave, and institutionalized care for the sick and aged—are the least religious. When the state is looking out for you, there's less need to seek help from above.

There is plenty of evidence that when people see themselves as less well off than others, or as in dismal situations or environments where they feel they have no control, they either become more religious or cling to their faith more tenaciously. A strong predictor of both religiosity and people's feeling of well-being is income inequality: even if you're relatively well off compared with other people in the world, you'll still feel marginalized if your countrymen are richer than you. Napoleon Bonaparte clearly saw the palliative effect of religion on this inequality, and its value in running a country: "I do not see in religion the mystery of the incarnation so much as the mystery of the social order. It introduces into the thought of heaven an idea of equalization, which saves the rich from being massacred by the poor."

In the United States, income inequality is one of the statistics most highly correlated with the national level of religious belief: the higher the inequality, the higher the average degree of religiosity. Tellingly, the two factors fluctuate in tandem, with religiosity increasing only after income inequality rises and decreasing only after it falls. This time delay, with the strength of religious belief changing *after* income inequality, and in the same direction, suggests that it's the inequality that breeds faith rather than the other way around.

Increasingly, then, we see that religious belief is a response to the uncertainties and hardships of life on this planet—a response to those difficulties, but not a way to lessen them. I am not a Marxist, but Marx got at least one thing right: for many, religion weakens the incentive to fix both personal and societal problems. And that is the biggest problem with seeing faith as a social palliative. People often criticize Marx for denigrating religion, citing his statement that "Religion . . . is the opium of the people." But seen in context, the quote is far more nuanced—a call for social change that would make religion superfluous:

> *Religious* distress is at the same time the *expression* of real distress and also the *protest* against real distress. Religion is the sigh of the oppressed creature, the heart of a heartless world, just as it is the spirit of spiritless conditions. It is the *opium* of the people.
>
> To abolish religion as the *illusory* happiness of the people is to demand their *real* happiness. The demand to give up illusions about the existing state of affairs is the *demand to give up a state of affairs which needs illusions.* The criticism of religion is therefore *in embryo the criticism of the vale of tears,* the *halo* of which is religion.

Can There Be Dialogue Between Science and Faith?

People regularly call for dialogue between science and religion, in which theologians, priests, and rabbis should sit down with scientists and hash out their differences. By "dialogue," the proponents don't just suggest that scientists and believers should talk to each other, but insist that such an exchange will dispel misunderstandings, benefiting both science and religion. Such conclaves are in fact held regularly, even at the Vatican. The motivation behind them is expressed in a famous quote from Albert Einstein: "Science without religion is lame, religion without science is blind." But that quote is torn from its context, where it's clear that what Einstein meant by "religion" is simply a deep awe before the puzzles of the universe. Einstein repeatedly denied the existence of a personal, theistic God, and saw Abrahamic faiths as fallacious and man-made institutions. He was at best a pantheist, viewing nature itself as "the divine." The blindness of science without religion refers to his belief that science goes nowhere without a profound and deep curiosity and wonder—traits that Einstein considered "religious." Einstein's views, often misconstrued, should give no solace to the majority of believers who are theists, nor to those who think that a science/faith dialogue would be mutually productive.

Nevertheless, is it possible to have a constructive dialogue? My response is that anything useful will come from a *monologue*—one in which science does all the talking and religion the listening. Further, the monologue will be constructive for only the listener. While scientists can learn more about the nature of belief by talking to the faithful, those benefits can accrue to anyone who wants to learn more about religion. In contrast, religion has nothing to tell scientists that can improve their trade. Indeed, the progress of science has required shedding all vestiges of religion, whether those be the beliefs themselves or religious methods for finding "truth." We do not need those hypotheses.

On the other hand, religion can benefit from science in several ways—if we conceive of "science" broadly and of "religion" as not just the beliefs but the institutions. First, science can tell us, at least in principle, about the evolutionary, cultural, and psychological basis of religious belief. There are many theories for why humans created religion, including fear of death, desire for a father figure, a need for social interaction, the wish of some people to control others, and the innate proclivity of humans to attribute natural events to conscious agents. It's my guess that, given the origins of religion in the distant and unrecoverable past, we'll never fully understand why and how it began. Nevertheless, we've seen new religions begin in recent years—Christian Science and Scientology are two—giving us the opportunity to study the psychological appeal of religion and perhaps the neurological correlates of belief.

Further, biblical scholarship, which when done properly is simply historical science applied to literature, can shed considerable light on the origins of scripture—light that can at least help the faithful make sense of their sacred books. We are now fairly sure, for instance, that the two creation stories in Genesis 1 and 2, which contradict each other about the origin of the Earth and its creatures, involve separate creation myths concocted several centuries apart.

Finally, what might be considered a real contribution of science to religious *belief* is the empirical demonstration that some of those beliefs are wrong. The many falsified biblical claims include the creation story, the claim of Adam and Eve as the ancestors of humanity, the Exodus of Jews from Egypt, and the census of Augustus that according to the Gospel of Luke brought Joseph and a pregnant Mary to Bethlehem. Because liberal and science-friendly faiths presumably don't want factually incorrect theologies, this forces them to turn what was previously taken literally into metaphors, and then into theological virtues. One can see these scientific corrections as "improving" faith, but only by removing the parts that are factually wrong.

Of course, it's useful for everyone, including scientists, to learn more about religion, for it's one of the driving impulses of humanity, directing the course of history (as in the present Middle East), profoundly affecting society (politics in modern America would be a mystery without understanding our hyperreligiosity), and contributing to the creation of great art, music, and literature. *Macbeth* is loaded with biblical allusions, and without a rudimentary knowledge of Christianity, Leonardo da Vinci's *Virgin of the Rocks* is simply a picture of a man, a woman, and two infants. But the historic and artistic importance of religion is not the point of religion/science dialogues, whose real aim is either to defend religion against science, to infuse science with religion, or to demonstrate that the two areas are valid and complementary ways of finding truth.

I have argued that religion is to science as superstition is to reason; indeed, that is the very reason they are incompatible. I've also maintained that this incompatibility rests on two pillars. One is that in some ways religion is like science, for most religions make claims about what exists in the universe, and purport to give evidence for those claims. (I emphasize again that there is far more to religion than truth claims!) And the value of a religion to its believers, regardless of what behavior it motivates, depends heavily on assenting to at least some of those claims. If Muslims knew that Muhammad, like Joseph Smith, was making up the words that became dogma; if Christians knew that Jesus was neither divine nor resurrected, but merely one of many apocalyptic preachers of that era; if theists knew that there were no documented interventions of God into the universe—then believers would melt away like spring snow. Yes, there are some sophisticated believers and theologians who see religion as independent of facts, but they are in the minority, and by and large their "religion"—often more a philosophy—does little harm to either science or society.

Irrationality enters when religion's truth claims are based not on reason or any kind of systematic investigation, but on *faith*—belief in matters for which there's no convincing evidence, but which are seen as true simply because people *want* them to be true, or were *taught* that they were true. This is, as Father Consolmagno said, thinking with the heart instead of the head. Coronary thinking is incapable of finding truth: millennia of religious conflicts and strife, resting on conflicting "truths" divined by faith, attest to that. In the Middle Ages theology was called "the Queen of the Sciences," but of course at that time "science" referred to any area of investigation. Nowadays, when we have real science, we realize that theology—the study of God, his nature, and his attributes—is as useless at understanding reality as when Thomas Paine characterized it in 1795:

> The study of theology, as it stands in Christian churches, is the study of nothing; it is founded on nothing; it rests on no principles; it proceeds by no authorities; it has no data; it can demonstrate nothing; and it admits of no conclusion. Not any thing can be studied as a science, without our being in possession of the principles upon which it is founded; and as this is not the case with Christian theology, it is therefore the study of nothing.

The second pillar of incompatibility is that the scientific pretenses of religion, when challenged, become pseudoscientific. That is, when the inevitable clashes ensue between brain- and heart-based thinking, religion resorts to the same pseudoscientific defenses used by Holocaust deniers, UFO devotees, and advocates of extrasensory perception. The vast majority of believers don't want their faith examined skeptically, nor do they honestly examine other faiths to find why they see their own as true and those others as false. Finally, like true pseudoscience, religion defends its claims by turning them into a watertight edifice immune to refutation. And what cannot be refuted cannot be accepted as true.

In the end, why isn't it better to find out how the world really works instead of making up stories about it, or accepting stories concocted centuries ago? And if we don't know the answers, why shouldn't we simply admit that we don't know, as scientists do regularly, and keep looking for answers using evidence and reason? Isn't it time that we take to heart the Apostle Paul's advice to the Corinthians to grow up and put away our childish things? Every obeisance we pay to faith buttresses those faiths that do real damage to our species and our planet.

It is time for us to stop seeing faith as a virtue, and to stop using the term "person of faith" as a compliment. After all, we don't call someone who believes in astrology, homeopathy, ESP, alien abduction, or even Scientology a "person of faith," even though that's precisely what such people are. The irony of extolling unfounded beliefs was expressed by Bertrand Russell, the most outspoken atheist of his time, in the very first sentence of his collection *Sceptical Essays:*

> I wish to propose for the reader's favourable consideration a doctrine which may, I fear, appear wildly paradoxical and subversive. The doctrine in question is this: that it is undesirable to believe a proposition when there is no ground whatever for supposing it true.

Or, as Sam Harris, Russell's modern counterpart, argues, "Pretending to be certain when one isn't—indeed, pretending to be certain about propositions for which no evidence is even conceivable—is both an intellectual and a moral failing."

Finally, although I'm a scientist, I am deeply moved by the wonders that science has brought to us in its short five centuries, and feel that religion is not only incompatible with science, but a roadblock to scientific progress, I am not proposing a robotic world governed by science. The world I want is one in which the strength of one's beliefs about matters of fact is proportional to the evidence. It is a world where it is okay to reserve judgment if one doesn't know the answer, and where it's not seen as offensive to doubt the claims of others.

A world that is faithless would not be without the arts, either. Those don't rest on faith, so imaginative art, literature, and music would still be with us. Too, we would retain justice, law, and compassion, perhaps in even greater measure than now, for our judgment wouldn't be warped by adherence to unevidenced divine strictures.

But wouldn't the end of faith also mean the end of morality and of the social benefits that come with religion? No, for the experience of Europe tells us this need not happen. Secular morality and nonreligious forms of communal experience are perfectly able to fill in the gaps when religion wanes. Indeed, secular morality, which is not twisted by adherence to the supposed commands of a god, is superior to most "religious" morality. And faith need not be replaced with other brands of faith: Europeans haven't shifted their belief in God to belief in ghosts and other paranormal phenomena. They've simply abandoned superstition altogether, and don't seem to need the "atheist churches" that are sprouting in the United States and United Kingdom.

I want to end with two stories about faith and science. The first involves Robert L. Park, a physics professor at the University of Maryland. On September 3, 2000, Park was on his customary jog through the woods when he had a gruesome accident. Its footing weakened by recent rains, a large oak tree beside the path fell on him as he was running by, crushing him and severely fracturing his arm and femur, which was driven right through the skin of his leg. He was pinned and unconscious. Fortunately, an undocumented immigrant from El Salvador found Park and called for help on his cell phone. Without the Salvadoran's phone, a product of science-based technology, Park, who was half a mile from the trailhead, would certainly have died. Two priests were also on the scene, but all they could offer was last rites.

Yet even when rescued, Park still would have died without modern medicine, especially antibiotics, which first appeared on the market in the mid-1940s. Because he had a large open wound through which soil bacteria entered his body, he required not only multiple surgeries and a temporary metal rod to bind his shattered femur, but also a catheter threaded through his arm into a vein near his heart, through which new and powerful antibiotics were infused into his body. It took nearly a year of this treatment before doctors beat back the infection.

Park's recovery required a concatenation of events, none of which could have occurred without science. As he said, "I was taken to the most modern hospital in the nation's capital, where I was put back together by skilled orthopedic surgeons guided by the latest medical imaging devices. They consulted frequently with various specialists. Psychiatrists monitored my emotional state; hematologists kept track of my blood tests, looking for indications of infection; caring professionals attended to me twenty-four hours a day; trained therapists guided me through rehabilitation."

The point is that as well-intentioned as believers may be (and the two priests later became Park's friends, strolling with him many times along the same trail), their faith is at best useless in such situations. There's little doubt that most people pinned under a tree would prefer to get medical help rather than prayer. Indeed, Park was an atheist, and had he been conscious he might well have been frightened by hearing incantations muttered by priests. But even had he been Catholic, there is no evidence that last rites would accomplish anything, for we can't be confident that even if there were a god, Catholicism, rather than, say, Islam, would be the one religion with the right god and beliefs. Faith has no way to find out. In the end, the priests' words were as meaningless and ineffectual a superstition as my avoiding the cracks in the sidewalk when I was a child. Many people would have called Park's recovery a miracle, but it wasn't: it was the result of decades of scientific inquiry in many areas, as well as the diligence of trained doctors, nurses, and rehabilitationists. No prayers or supernatural interventions were involved.

For many, the transition from religion to nonbelief, from faith to rationality, is an awakening. Although that awakening may bring a sense of freedom and self-determination, it is sometimes catalyzed by tragedy and accompanied by regret for a life spent in servitude to superstition. Such was the case of Russ Briggs.

Briggs was a member of the Followers of Christ Church in Oregon, a sect that rejects medical care. He and his wife lost two boys shortly after birth, within a year of each other. One was premature but could easily have been saved had he not been tended by an untrained midwife. After the deaths, Briggs left the church in 1981. When trying to rejoin later, he was rebuffed and ostracized by his family and other Followers, one of whom publicly called him "a liar and a whoremonger." Tormented by guilt, he continued to visit his sons' graves, vividly describing his pain: "I stood there, a twenty-year-old child, sobbing and hurting, and trying to figure out why my child died. Had there been an incubator there, or modern medicine, I know he would have made it." He added, "I could have saved them, but I let them die." This was an act of courage, for Briggs took full responsibility for what he did, avoiding rationalizing or covering his acts with the blanket of excuses called "God's will." He stands alone, having shed superstition and accepted the reality of what happened.

Briggs later accepted conventional medical care and produced two healthy daughters. Let those who see faith as a virtue—an attitude that allows the Followers to escape legal responsibility for child abuse—ponder what Briggs said when recalling how his religion killed his sons: "It's only when you no longer have that belief that all of a sudden it comes to you: How could I ever have done that?"

ACKNOWLEDGMENTS

As this book is only tangentially connected with my day-to-day work in evolutionary genetics, I have benefited greatly from the help and encouragement of diverse friends and colleagues, including Dan Barker, Andrew Berry, Russell Blackford, Paul Bloom, Peter Boghossian, Maarten Boudry, Sarah Brosnan, Sean Carroll, Matthew Cobb, Graham Coop, Martin Corcoran, Richard Dawkins, Dan Dennett, Michael Fisher, Yonatan Fishman, Faye Flam, Caroline Fraser, Karl Giberson, Anthony Grayling, Miranda Hale, Larry Hamelin, Sam Harris, Will Hausman, Alex Lickerman, John Loftus, Eric MacDonald, Anne Magurran, Peggy Mason, Greg Mayer, Steve Pinker, Leslie Rissler, Jason Rosenhouse, Allen Sanderson, Michael Shermer, Grania Spingies, the late Victor Stenger, Sue Strandberg, and Ed Suominen. Hugh Dominic Stiles was indispensable for finding the source of many obscure quotations. Not all of these people, of course, will agree with everything I've written, and I apologize to those whose names have been inadvertently omitted. Many of the ideas and themes in this book were developed in posts on my Web site, whyevolutionistrue .wordpress.com, and I am grateful to the dozens of readers whose comments contributed to my own thinking. Finally, I benefited once again from the advice and help of my agent, John Brockman, and from the astute editorial skills of Wendy Wolf at Penguin Random House.

Parts of chapter 3 are modified from articles in the *New Republic* and the *Times Literary Supplement* (Coyne 2000, 2009b), while parts of chapter 4, on religious critiques of science, are modified from pieces originally published on my Web site (Coyne 2013a, 2013b) and in *Slate* (Coyne 2013c).

NOTES

All quotations from the Bible are taken from the King James Version, and definitions from the *Oxford English Dictionary* were accessed from the electronic version available at the University of Chicago Libraries. When using the word "god," I capitalize it when it refers to the Abrahamic God of Islam, Judaism, and Christianity, but leave it lowercase when it refers to generic gods. Because the Abrahamic God is conventionally referred to as "he," I use that word as well but have not capitalized it.

Epigraphs

vii **"God is an hypothesis":** Shelley 1915 [1813], p. 5.
vii **"We have already compared":** Ingersoll 1900a, pp. 133–34.

Preface

xi **"The good thing about science":** "The good thing about science . . . Neil deGrasse Tyson," YouTube, https://www.youtube.com/watch?v=yRxx8pen6JY.
xiii **But my vague beliefs in a God:** Manier 2008.
xiv **the proportion of creationists:** My book was *Why Evolution Is True* (Coyne 2009a); the Gallup poll on evolution is Gallup 2014.
xxi **"The point is not that we atheists can prove":** Harris 2007.
xxii *Life After Faith:* Kitcher 2014.
xxii **Pascal Boyer's *Religion Explained* and Daniel Dennett's *Breaking the Spell*:** Boyer 2002; Dennett 2006.

Chapter 1: The Problem

1 **"For we often talked of my daughter":** Tennyson, "The Village Wife," http://www.telelib.com/authors/TennysonAlfred/verse/ballads/villagewife.htm/. The original was:

Fur hoffens we talkt o' my darter es died o' the fever at fall:
An' I thowt 'twur the will o' the Lord, but Miss Annie she said it wur draäins.

3 **"Then has it in truth come to this":** Draper 1875, p. 363.

3 **"persons of every religious denomination":** Cornell University 1892.

3 **"So far from wishing to injure Christianity":** A. D. White 1932, p. vi.

4 **"Then it was that there was borne":** Ibid., p. viii.

5 **And while not all of the "science and religion" books:** WorldCat search, January 20, 2014. Between 1974 and 1983, WorldCat lists 48,577 books published in English on the topic of religion. Of these, 514, or 1.06 percent, were on "science and religion." In the next decade, 1984–93, the proportion remained similar: 0.96 percent (606 out of 63,120). But over the last two decades the proportion nearly doubled: 1.40 percent from 1994 to 2003 (1,274 out of 90,906) and 2.33 percent between 2004 and 2013 (2,574 out of 110,259).

6 **"By one report, U.S. higher education":** Larson and Witham 1997, p. 89.

6 **"building bridges between science and theology":** Center for Theology and the Natural Sciences, http://www.ctns.org/index.html.

7 **Dialogue on Science, Ethics, and Religion:** http://www.aaas.org/DoSER.

8 **the John Templeton Foundation:** Bains 2011.

8 **the Clergy Letter Project:** http://www.theclergyletterproject.org.

8 **"The sponsors of many of these":** American Association for the Advancement of Science 2006.

8 **But because many Americans believe otherwise:** Statistics on young-Earth creationism from Gallup 2014.

9 **"The science of evolution does not make claims":** Hess 2009.

9 **Because nearly 20 percent of Americans:** American acceptance of evolution from Gallup 2014; proportion of American nonbelievers from Pew Research 2012a.

9 **"Though faith is above reason":** United States Catholic Conference 1994, Section 159.

10 **Further, as we'll see:** Masci 2009.

10 **When asked what they would do:** *Time*/Roper poll cited in Masci 2007; D. Masci, personal communication.

10 **A related poll:** Gallup 2007.

11 **A 2009 Pew poll:** Pew Research 2009a; 68 percent of those *not* affiliated with a church saw a conflict between science and religion.

11 **"Reason #3":** Barna Research 2011.

11 **One of the more remarkable demonstrations:** Citizens for Objective Public Education 2013.

12 **Surveying American scientists:** Larson and Witham 1997; Pew Research 2009b.

12 **When one moves to scientists:** Ecklund 2010.

12 **Finally, sitting at the top tier:** Larson and Witham 1997.

13 **So, among academics:** Gross and Simmons 2007, 2009.

13 **The first is that elite scientists:** Ecklund and Scheitle 2007.

13 **But there's further evidence:** Decline of religious belief in America: Grant 2008, 2014. Older scientists being less religious: Pew Research 2009b; Gallup 2011. Scientists between eighteen and thirty-four years old, for instance, are significantly more likely to believe in God (42 percent) than are scientists over sixty-five (28 percent). Conversely, scientists who reject both God and a "higher power" are more frequent in the older group (48 percent) than in the younger (32 percent).

15 **"Despite a million chances":** Coyne 2009a, p. 223.

16 **This makes "nones" the fastest-growing category:** Pew Research 2012a.

17 **"Sir John believed"**: John Templeton Foundation, "Philanthropic Vision," http://www
   .templeton.org/sir-john-templeton/philanthropic-vision.
17 **"Sir John's own eclectic list"**: John Templeton Foundation, "Science and the Big Questions," http://www.templeton.org/what-we-fund/core-funding-areas/science-and-the
   -big-questions.
18 **You may have encountered the foundation:** John Templeton Foundation, "Big Questions Essay Series," http://www.templeton.org/signature-programs/big-questions
   -essay-series.
18 **The foundation's most famous award:** Bains 2011; John Templeton Foundation,
   "About the Prize" at http://www.templetonprize.org/abouttheprize.html.
19 **"The Foundational Questions"**: John Templeton Foundation, "Foundational Questions in Evolutionary Biology (FQEB)," http://www.templeton.org/what-we-fund
   /grants/foundational-questions-in-evolutionary-biology-fqeb.
20 **the $100,000 Epiphany Prize:** Epiphany Prizes, http://www.epiphanyprizes.com.
20 **the World Science Festival:** John Templeton Foundation, "The World Science Festival:
   Big Ideas Series," http://www.templeton.org/what-we-fund/grants/the-world-science
   -festival-big-ideas-series.
20 **Important New Atheist works:** Harris 2004, 2006; Dawkins 2006; Dennett 2006;
   Hitchens 2007a; Stenger 2007.
22 **"I am the way"**: John 14:6.
23 **"extraordinary claims require"**: Hitchens 2003.
23 **"First, we hypothesize"**: Hamelin 2014.

Chapter 2: What's Incompatible?

26 **"I admit I'm surprised"**: Angier 2004, pp. 132–33.
28 **"a testable body of knowledge"**: Shermer 2013, p. 208.
28 **"The first principle is that you must not fool yourself"**: Feynman and Leighton
   1985, p. 343.
29 **"The interest I have in believing"**: Voltaire and Arouet 1763, p. 10: "De plus, l'intérêt
   que j'ai à croire une chose n'est pas une preuve de l'existence de cette chose."
30 **"What distinguishes knowledge"**: Kaufmann 1958, p. 78.
31 **"confirmed to such a degree"**: Gould 1983, p. 255.
31 **But some people take this too far:** See Lehrer 2010; Flam 2014; Johnson 2014.
38 **"I can live with doubt"**: Richard Feynman, BBC *Horizon* interview, 1981, transcribed
   from video at https://www.youtube.com/watch?v=I1tKEvN3DF0.
38 **"the dog sniffing tremendously"**: Mencken 1922, p. 270.
39 **The participants in the discovery:** Tunggal 2013.
40 **"As I prepared to leave Little Rock"**: Gould 1982, p. 17.
41 **about 54 percent of the word's inhabitants:** http://www.adherents.com/Religions
   _By_Adherents.html.
42 **"[T]here is a certain uniform deliverance"**: James 1928, p. 508.
43 **We all know Catholics:** Rejection of evolution by American Catholics in Masci 2009.
44 **"Religion isn't a philosophical argument"**: Spufford 2012, pp. 34–35.
44 **"It is a shame that this word"**: Aslan 2005, p. xviii.
45 **"Now if Christ be preached"**: 1 Corinthians 15:12–14.
45 **"For the practices of the Christian religion"**: Swinburne 2012, p. 120.
45 **"A religion therefore contains"**: Stenmark 2012, p. 65 (emphasis in the original).
46 **"The question of truth"**: Polkinghorne 2011, p. 2.
46 **"A religious tradition"**: Barbour 2000, pp. 36–37.

46 **"Likewise, religion in almost all of its manifestations":** Giberson and Collins 2011, p. 86.

47 **A 2011 survey of belief:** Ipsos/Reuters 2011.

47 **But three surveys:** Smith 2012.

47 **"The Untold Story of Creation":** *Awake!*, March 2014, p. 4, http://www.jw.org/en /publications/magazines/g201403/untold-story-of-creation/.

48 **"Attributes of God":** Franciscan Clerics of Holy Name College 1943, pp. 147–48.

48 **"I take the proposition":** Swinburne 2004, p. 7.

48 **"What [Daniel Dennett] calls":** Plantinga 2011, p. 11.

49 **"It's really important":** Bickel and Jantz 1996, p. 40.

49 **Liberal theologians like Karen Armstrong and David Bentley Hart:** Armstrong 2009; Hart 2013.

51 **The most recent survey:** Harris Interactive 2013.

52 **To get data on the content:** Baggini 2011 (includes both survey results and quotation).

52 **The 2011 Ipsos/Reuters poll:** Ipsos/Reuters 2011.

53 **The world's Muslims:** Pew Research 2012b.

54 **"Only a small minority of believers":** Wieseltier 2013.

54 **"non-negotiables of Christianity":** Dembski 2012.

55 **"There's no evidence":** Sullivan 2011, responding to Coyne 2011.

56 **"When, however, there is question":** Pius XII 1950, para. 37.

56 **"Regardless of how differently":** Livingstone 2011, p. 5. My thanks to Jason Rosenhouse for pointing out this quote.

57 **"Augustine says . . . 'Three general opinions prevail'":** Aquinas, *Summa Theologica*, Question 102, Article 1, http://www.newadvent.org/summa/1102.htm#article1.

58 **"the tragedy of theology":** Bernstein 2006, p. 26.

58 **"The narrative indeed":** Augustine 2002, pp. 346–47.

59 **Julian Baggini's online survey:** Baggini 2011.

59 **And yet it is supported:** Coyne 2009a.

59 **And indeed, evolution:** Worldwide and U.S. surveys on evolution from Ipsos/Reuters 2011; Coyne 2012; Gallup 2014.

60 **When asked in 2007:** Gallup 2007.

60 **This arrant rejection of facts:** Kahan 2014 and references therein.

61 **Nevertheless, 27 percent:** John Paul II 1996; Masci 2009.

61 **A survey of Americans:** Pew Research 2010.

62 **The psychologist Jon Haidt:** Haidt 2013.

62 **How many Christians:** See Ehrman 2013 on the historicity of Jesus.

62 **"People come to faith":** Luhrmann 2012, p. 223.

62 **"The aim is to discover":** Plantinga 2011, p. 154.

63 **"[In Christianity, Judaism, and Islam]":** Plantinga 2010.

67 **"the substance of things hoped for":** Hebrews 11:1.

67 **"intense, usually confident, belief":** Kaufmann 1961, p. 2.

68 **"blessed are they that have not seen":** John 20:29.

68 **"But the natural man receiveth":** 1 Corinthians 2:14.

68 **"The Son of God died":** Evans 1956 (". . . et mortuus est Dei filius: prorsus credibile est, quia ineptum est; et sepultus resurrexit: certum est, quia impossibile").

68 **"To be right in everything":** Fordham University, "Medieval Sourcebook: St. Ignatius Loyola: Spiritual Exercises," Thirteenth Rule, http://www.fordham.edu/halsall/source /loyola-spirex.asp.

68 **"The spirit of curiosity is not a good spirit":** Elise Harris, "Pope Francis: Kingdom

of God 'Comes by Wisdom,'" Aleteia.org, November 14, 2013, http://www.aleteia
.org/en/religion/news/pope-francis-kingdom-of-god-comes-by-wisdom-14774001.

69 **"For reason is the greatest enemy":** Luther 1857, p. 164.

69 **"There is on earth among all dangers":** Luther quoted in Kaufmann 1961, p. 75.

70 **The Lutheran theologian whom I debated:** Lutheran creeds from Evangelical Lutheran Church in America, http://www.elca.org/Faith/ELCA-Teaching.

70 **"an expression of the determination":** Royal Society, "History," http://royalsociety
.org/about-us/history/.

72 **"Natural selection was a revolution":** Atkins 1995, p. 98.

73 **"And therefore, if in some historically contingent circumstances":** William Lane
Craig, "Dealing with Doubt," video at http://www.youtube.com/watch?v=S-fDyPU
3wlQ; see also "Set Free" at http://www.jcnot4me.com/Items/contra_craig/contra
_craig.htm#Comments%20on%20Craig%27s%20Book:%20Reas.

73 **"Let's take the resurrection":** Justin Thacker in interview with Julian Baggini, in
Baggini 2012a, pp. 516–17.

73 **"if you ask me":** John Haught in interview with Steve Paulson, at http://asylum.za.org
/viewtopic.php?t=1860&sid=e5bccce6ea47858d464185b968a14142; see also Paulson
2010, pp. 83–98.

74 **"As a believer in God":** Giberson 2009, p. 213.

75 **"Genesis is a book of religious revelations":** Ayala 2007, p. 175.

75 **"As we have seen":** Gilkey 1985, pp. 113–14.

75 **"How should we understand the narratives of Genesis?":** Pope Benedict, 2013. Text
from http://en.radiovaticana.va/news/2013/02/06/audience:_god,_creation_and_free
_will/en1-662454.

75 **Unfortunately, many believers:** Jones 2011.

76 **"The Bible's prime purpose":** Wieland 2000, p. 4.

78 **"A world of life with evolution":** Ayala 2012, p. 291.

78 **"Seeking to populate this otherwise sterile universe":** Collins 2006, pp. 200–201.

79 **"The presence of God is veiled":** Polkinghorne and Beale 2009, p. 11.

79 **"It is essential to religious experience":** Haught 2003, p. 86.

79 **"The other day I was praying":** Weaver 2014.

80 **"The invisible and the non-existent":** McKown 1993, p. 39.

80 **Given that nobody has returned:** For criticisms of two recent reports of visits to
heaven, see Dittrich 2013 and T. A. McMahon, "'Is Heaven for Real' for Real? An
Exercise in Discernment," at *The Berean Call* (a religious site), http://www.theberean
call.org/content/heaven-real-real-exercise-discernment-0/.

80 **"In any case, were I to try":** Haught 2006, pp. 203–4.

81 **"With respect to the theological view":** Darwin letter to Asa Gray, May 22, 1860. Darwin Correspondence Project, letter 2814, http://www.darwinproject.ac.uk/entry-2814.

83 **If born in Utah:** Newport 2014a.

84 **The Quran also claims:** Quran, Surat An-Nisā' 4:157.

84 **How many different religions are there?:** Number of Christian sects from Gordon
Conwell Theological Seminary, http://www.gordonconwell.edu/resources/documents
/StatusOfGlobalMission.pdf.

85 **And even today:** Pew Research 2011.

85 **"It is *highly* likely":** Loftus 2013, pp. 16–17.

86 **"argument from symmetry":** Kitcher 2014, chapter 1.

86 **The progress of science:** Increase in life span from Finch 2010.

88 **"The nesting relationship of successive scientific theories":** Polkinghorne 1994,
pp. 7–8.

89 **"The slow progress in philosophy and theology"**: Moreland 1989, pp. 238–39.
91 **"Name me an ethical statement"**: Hitchens 2007b, p. xv. Hitchens's corollary to this question is "If you ask an audience to name a wicked statement or action directly attributable to religious faith, nobody has any difficulty finding an example."
93 **"It is not that the methods and institutions of science"**: Lewontin 1997, p. 31.
93 **"First, science is a limited way of knowing"**: Scott 1996, p. 518.
94 **"Taken together, the (1) proven success"**: Forrest 2000, p. 21.

Chapter 3: Why Accommodationism Fails

97 **"There is no harmony between religion and science"**: R. G. Ingersoll, "The Truth Seeker," September 5, 1885, http://www.gutenberg.org/files/38808/old/orig38808-h /main.htm.
97 **"Cognitive dissonance" is a well-known phenomenon**: Tavris and Aronson 2007, especially chapter 1.
98 **Some of these, whom I call "faitheists"**: "Belief in belief" discussed in Dennett 2006, chapter 8.
98 **"We have a political battle"**: My post: http://whyevolutionistrue.wordpress.com /2011/04/21/another-tom-johnson-did-dawkins-call-religious-people-nazis/; Stanyard's comment: http://whyevolutionistrue.wordpress.com/2011/04/21/another-tom-johnson -did-dawkins-call-religious-people-nazis/#comment-94522.
99 **"Unless at least half of my colleagues"**: Gould 1987, p. 70.
100 **Even Indian scientists**: Ram 2013.
100 **In America, religious scientists**: Ecklund 2010.
101 **"The most beautiful and deepest experience"**: Albert Einstein, "My Credo," 1932, *Albert Einstein in the World Wide Web*, http://www.einstein-website.de/z_biography/credo.html.
101 **"I believe in Spinoza's God"**: Quoted in Isaacson 2007, pp. 388–89.
102 **"dilutes religion to the point of meaninglessness"**: Miller 1999, p. 221.
102 **"Gingerich is still baffled"**: Tom Mueller, "Whale Evolution," *National Geographic*, August 2010, http://ngm.nationalgeographic.com/2010/08/whale-evolution/mueller-text.
103 **"frisson in the breast"**: Dawkins described his spirituality in an interview on Al Jazeera, quoted in Mooney 2010.
103 **"Our results show unexpectedly"**: Ecklund and Long 2011, pp. 255, 261–62.
103 **"truth cannot contradict truth"**: Address of Pope John Paul II to the Pontifical Academy of Sciences, October 22, 1996, http://www.newadvent.org/library/docs_jp02tc.htm.
103 **"God is the source of both reason and revelation"**: Pope 2004, p. 189.
105 **"The fire you kindle arises from green trees"**: Nurbaki 2007, p. 133, translation of Quranic verse 36:80.
105 **"simply grabs whatever theory"**: Nanda 2004.
106 **"If scientific analysis were conclusively to demonstrate"**: Dalai Lama 2005, p. 3.
106 **"Remember the widely different aspects"**: Whitehead 1925, p. 265.
107 **"that grants dignity and distinction"**: Gould 1999, p. 51.
107 **"Science tries to document"**: Ibid., p. 4.
108 **"the potential harmony"**: Ibid., p. 43.
109 **"Religion just can't be equated"**: Ibid., pp. 209–10.
109 **"Thou shalt not mix the magisteria"**: Ibid., pp. 84–85.
109 **"In other words"**: Ibid., pp. 128, 148–49.
110 **Recall that 42 percent of Americans**: Polls on belief in and teaching of evolution: Gallup 2014; Newport 2014b.
110 **"quibble about the labels"**: Gould 1999, p. 62.
111 **"But the religion"**: Hutchinson 2011, p. 207.

111 **"[A] closer look at Gould's writings":** Haught 2003, pp. 6–7.
112 **"Because they are not a part of nature":** National Academies 2008, p. 1.
112 **"Science is a method":** National Science Teachers Association 2013.
113 **"[if] we could apply natural knowledge":** Pennock 1999, p. 290.
114 **"1. Intercessory prayer can heal the sick":** Fishman and Boudry 2013, p. 929.
115 **It's surprising how often Americans pray:** Gallup and Lindsay 1999, p. 46; Jones 2010; Luhrmann 2012; U.S. News and Beliefnet Prayer Survey: http://www.beliefnet.com/Faiths/Faith-Tools/Meditation/2004/12/U-S-News-Beliefnet-Prayer-Survey-Results.aspx.
115 **Over 35 percent of Americans . . . 24 percent:** Barnes et al. 2004; McCaffrey et al. 2004.
115 **In fact, such a test:** Galton 1872.
115 **there are more modern and scientifically controlled studies:** Aviles et al. 2001; Krucoff et al. 2005; Astin et al. 2006; Benson et al. 2006; Schlitz et al. 2012.
116 **There's no substantive difference:** Offit 2012.
116 **The "miracle" that clinched:** Rohde 2003.
117 **"When I was at Lourdes":** France 1894, p. 203 (my translation).
117 **The question, "Why won't God heal amputees?":** See "Why Won't God Heal Amputees?" at http://whywontgodhealamputees.com/.
117 **"That sensible":** France 1894, p. 204 (my translation).
118 **In his book *The Varieties of Scientific Experience:*** Sagan 2006, chapter 6.
118 **"Your question what would convince me":** Darwin letter to Asa Gray, September 17, 1861. Darwin Correspondence Project, letter 3256, http://www.darwinproject.ac.uk/letter/entry-3256.
119 **"Any sufficiently advanced technology":** Clarke 1973, p. 21.
120 **"As a purely practical matter":** Giberson 2009, pp. 155–56.
120 **"a religious faith that depended upon a belief":** Hutchinson 2011, p. 222.
121 **The classic test for the truth of miracles:** Hume 1975 [1748].
122 **The Shroud of Turin:** Philip Pullella, "Italian Scientist Reproduces Shroud of Turin," Reuters, October 5, 2009, http://www.reuters.com/article/2009/10/05/us-italy-shroud-idUSTRE5943HL20091005; Inés San Martin, "Pope Francis to Venerate Famed Shroud of Turin," *Crux*, November 5, 2014, http://www.cruxnow.com/church/2014/11/05/pope-francis-to-venerate-famed-shroud-of-turin-in-2015/.
122 **In 2012, a statue of Jesus:** J. White 2012.
123 **One was suggested by the philosopher Herman Philipse:** Philipse 2012, chapter 10.
123 **"The body of Jesus probably decayed":** Funk and the Jesus Seminar 1998, p. 36.
124 **Of course, more conservative Christians:** Branch 1995.
125 **"God created man in his own image":** Genesis 1:27.
126 **"For since by man came death":** 1 Corinthians 15:21–22; see also Romans 5:12–21.
126 **As we've learned, Augustine:** Augustine, *City of God,* Book XIV, http://www.newadvent.org/fathers/120114.htm.
126 **"The account of the fall":** United States Catholic Conference 1994, Section 2, chapter 1, paragraph 7.
126 **And Americans as a whole:** Bishop et al. 2010.
127 **The genetic evidence tells us:** Y chromosome data from Francalacci et al. 2013; Poznik et al. 2013; Mendez et al. 2013. Other genome data from Garrigan and Hamer 2006.
127 **But the evidence is even stronger:** Henn, Cavalli-Sforza, and Feldman 2012; Li and Durbin 2011; Sheehan, Harris, and Song 2013.
128 **"The denial of an historical Adam and Eve":** Mohler 2011.
128 **"Really, without a doctrine of original sin":** Aus 2012.
128 **"For the faithful cannot embrace":** Pius XII 1950; see also United States Catholic Conference 1994, Section 390.

129 **"According to this model":** Alexander 2010–11.
130 **The most "sophisticated" attempt:** Enns 2012.
130 **"One can believe that Paul is correct":** Ibid., p. 143.
130 **"Paul's handling of Adam":** Ibid., p. 102.
131 **"possess[ed] this land among themselves":** Book of Mormon, 2 Nephi 1:5–9; see also Ether 13:2.
131 **But as with the existence of Adam and Eve:** Genetic data on Native Americans from N. A. Rosenberg et al. 2002; Murphy 2002.
132 **"DNA studies cannot be used":** Church of Jesus Christ of Latter-day Saints, "Book of Mormon and DNA Studies," 2014, https://www.lds.org/topics/book-of-mormon-and -dna-studies.
132 **On twelve occasions since 1982:** Gallup 2014; Newport 2014b.
133 **"The Teaching Authority of the Church":** Pius XII 1950, Section 36.
133 **And because life itself:** Pross 2012.
134 **"in [a modern human's] large and well-developed brain":** Wallace 1870, p. 343.
135 **"Religions can put up with all kinds":** Haught 2003, p. 185.
135 **"What is *not* consistent with Christian belief":** Plantinga in Dennett and Plantinga 2010, pp. 4–5.
136 **"The world of molecules":** Barbour 2000, p. 164.
136 **While ID arguments:** See Pennock 1999; Miller 1999, 2008; Coyne 2005; Orr 2005.
136 **"With man, we find ourselves":** John Paul II 1996.
137 **"Fortunately, in scientific terms":** Miller 1999, p. 241.
138 **What we *have* shown:** Sniegowski 1995.
143 **"Contrary to popular belief":** Conway Morris 2003, p. xv.
143 **"But as life re-explored adaptive space":** Miller 2008, pp. 152–53.
144 **In his book *Wonderful Life:*** Gould 1989.
145 **"a wildly improbable evolutionary event" and "a cosmic accident":** Ibid., pp. 44, 291.
147 **"The idea that secondary causes":** Haught 2003, p. 57.
148 **"The theory of evolution provided":** Ayala 2007, pp. 4–5.
148 **"But any world that contains atonement":** Plantinga 2011, p. 59.
149 **"arbitrary superfluity":** Grayling 2013.

Chapter 4: Faith Strikes Back

151 **"When I was working as a pastor":** Aus 2012.
152 **"there must always be many extraordinary facts":** Humphrey 1996, p. 71.
152 **"No one infers a god":** Ingersoll 1879, p. 56.
153 **"the attempt to argue for the truth":** Philipse 2012, p. 14.
154 **"As far as the examination of the instrument goes":** Paley 1809, p. 18.
155 **"Men think epilepsy divine":** Hippocrates of Cos, ca. 400 BCE. Quoted in Sagan 1996, p. 8.
155 **"How wrong it is to use God as a stop-gap":** Bonhoeffer 1967, p. 311.
155 **"A word of caution is needed":** Collins 2006, pp. 92–93.
157 **As for the origin of life:** Hazen 2005; Pross 2012; Keller, Turchyn, and Ralser 2014.
158 **Neuroscience has already made:** Baars, Banks, and Newman 2003; Pinker 2007 and personal communication; Dehaene and Changeux 2011.
159 **"the unreasonable effectiveness of mathematics in the natural sciences":** Wigner 1960.
160 **A full 69 percent of Americans:** Bishop et al. 2010.
160 **"It almost seems":** Miller 1999, pp. 228, 232.

163 **Assuming that life can inhabit:** Siegel 2013.
165 **"What reasons can God have had":** Philipse 2012, p. 276.
165 **"But why should we think of God":** Craig 2012, responding to Carroll 2012.
166 **"The following proposition":** Darwin 1871, pp. 71–72.
167 **"Scientists and humanists":** Wilson 1975, p. 562.
167 **"the sudden appearance in consciousness":** Haidt 2001, p. 818.
168 **"But humans are unique":** Collins 2006, pp. 27, 140.
168 **"Selfless altruism . . . is quite frankly a scandal":** Ibid., p. 200.
168 **"There are specific human experiences":** Linker 2014.
169 **In his book *The Blank Slate:*** Pinker 2002, citing D. Brown 1991.
169 **A survey of sixty "traditional societies":** Curry, Mullins, and Whitehouse in preparation.
169 **There are also moral instincts:** Thompson 1976; Hauser 2006.
170 **In *The Better Angels of Our Nature:*** Pinker 2011.
171 **The primatologist Frans de Waal:** de Waal 2006, 2013; de Waal, Machedo, and Ober 2009.
171 **capuchin monkeys seem to show:** See the video at http://www.youtube.com /watch?v=-KSryJXDpZo.
171 **Indeed, even rats:** Bartal, Decety, and Mason 2011; Bartal et al. 2014.
171 **When you overlay animal "morality":** Brosnan 2011, 2013.
172 **The work of the child psychologist Paul Bloom:** Bloom 2013.
172 **"There is no support":** Bloom 2014.
173 **In fact, many aspects:** Price 2012.
174 **As Peter Singer explains:** Singer 1981.
174 **Explaining altruism toward relatives:** See Dawkins 1976 and references therein.
175 **I've just seen a video:** See https://www.youtube.com/watch?v=570khFoaE4s.
177 **"What evolution underwrites":** Plantinga 2011, p. 316.
178 **"God created both us and our world":** Ibid., p. 269.
178 **"This capacity for knowledge of God":** Plantinga 2000, p. 180.
178 **"What I'll argue":** Plantinga in Dennett and Plantinga 2010, p. 17.
179 **"There is superficial conflict":** Plantinga 2011, p. ix.
179 **"Members of our species commonly believe":** Pinker 2005, p. 18.
180 **"checker-shadow illusion":** The illusion can be seen at http://www.michaelbach.de /ot/lum_adelsonCheckShadow/.
180 **The skeptic and science writer Michael Shermer:** Shermer 2002, 2013.
181 **"Were it not for sin and its effects":** Plantinga 2000, p. 214.
182 **The anthropologist Pascal Boyer:** Boyer 2002; for the development of religious beliefs over time, see Banerjee and Bloom 2013.
182 **Rather, religion is likely to be:** See Gould and Vrba 1982 for a discussion of exaptations.
183 **"We see, then, that whether we compare":** Wallace 1870, p. 343.
184 **"These abilities far surpass":** Plantinga 2011, p. 286.
185 **"All knowledge that is not the genuine product of observation":** Lamarck 1820, p. 84 (my translation): "Toute connaissance qui n'est pas le produit réel de l'observation ou de conséquences tirées de l'observation, est tout-à-fait sans fondement, et véritablement illusoire."
185 **"science is not the only way of knowing":** Collins 2006, p. 229.
186 **"confirmed to such a degree":** Gould 1983, p. 255.
188 **Nevertheless, microeconomics has produced:** Chetty 2013.
188 **The physicist Sean Carroll:** Carroll 2013.

189 **"Philosophy of science":** See, for instance, http://evolvingthoughts.net/2010/06/evolution-quotes-feyman-on-religion-and-science.

189 **Philosophy, for instance:** For example, Dennett 1991, 1995; Kitcher 1982, 1985; Pennock 1999.

189 **The Old Testament God:** See Deuteronomy 20:17–18 and 1 Samuel 15:3.

190 **Sam Harris, for instance:** Harris 2011.

191 **"Over a century after publication":** Charlton and Verghese 2010, pp. 94–95.

192 **"Consider the kind of putative insights":** Kieran 2009, pp. 194–95.

194 **"We know the ways in which humans express":** Comment by "Vaal," May 8, 2012, http://whyevolutionistrue.wordpress.com/2012/05/08/quote-of-the-day-mike-aus/#comment-216549.

196 **"It is not merely the misapplication":** Hutchinson 2011, p. 175.

196 **"What I meant by 'scientism'":** Haack 2012, p. 76.

197 **"But beneath the weighty ethical concerns":** Kass 2007, quoted in Pinker 2013, p. 30.

197 **"'Scientism' is a term of abuse":** Noordhof 1995.

198 **A few academics:** Wilson 1998; A. Rosenberg 2012.

198 **"History, for example":** Baggini 2012b.

199 **"is not an imperialistic drive":** Pinker 2013; for a critical response see Wieseltier 2013.

200 **"Consider the beauty of a sunset":** Hutchinson 2011, p. 54.

201 **"is a completely undefined term":** Paulson 2010, p. 171.

201 **"Of course an atheist can't prove":** Jacoby 2013.

202 **"The issue of God":** Kenneth Miller, interview on *Today Programme*, BBC, April 29, 2009, http://news.bbc.co.uk/today/hi/today/newsid_8023000/8023996.stm.

203 **"Once again, the only sensible approach":** Sagan 1996, p. 173.

204 **"Finally, we note":** Giberson and Collins 2011, p. 109.

205 **"Clearly, then, both religion":** Davies 2007.

205 **"If you find the idea":** Sarewitz 2012.

206 **"I'm not a biologist":** Redlawsk 2013.

206 **"There's the deeper worldview":** Haught 2007.

206 **Many older readers:** Finch 2010.

210 **"There is a very, very important difference":** Dawkins 1997.

211 **Isn't science, as some maintain:** Cortical Rider 2012.

212 **"The rise of science":** Stark 2005, pp. 22–23.

212 **"The very notion of physical law":** Davies 2007.

212 **"The ethical and moral acceptability":** Hutchinson 2011, p. 224.

213 **"My explanation . . . is":** Whitehead 1925, p. 19.

213 **"Any list of the giants of physical science":** Hutchinson 2003, p. 75.

214 **As the historian Richard Carrier has argued:** Carrier 2010.

214 **The historians Richard Carrier, Toby Huff, Charles Freeman, and Andrew Bernstein:** Freeman 2003; Huff 2003; Bernstein 2006.

214 **"In the Middle Ages, the great minds":** Bernstein 2006, p. 26 (reviewing Stark 2005).

217 **"Science is a revolutionary activity":** Miller 2012.

217 **"Critics of religion":** Small 2011.

218 **"It's not a charge I'd throw around":** Cohen 2014.

219 **is like blaming architecture:** Analogy from Pinker 2013.

220 **"With or without religion":** Weinberg 1999, p. 48.

221 **"For a successful technology":** Feynman 1986.

221 **But Weinberg *was* on the money:** Weinberg 1999, p. 207; Popper quote from Popper 1957, p. 244.

222 **"We must also be careful":** Small 2011.

222 **"Many other historical blunders":** Texas Charter School statement, in Scaramanga 2013.

224 **the development of "golden rice":** Golden Rice Project, http://www.goldenrice
.org/Content3-Why/why1_vad.php.

Chapter 5: Why Does It Matter?

225 **"A surgeon once called upon a poor cripple":** Ingersoll 1900b, p. 87.
226 **"empty of any claims":** Hutchinson 2011, p. 207.
228 **"The Foundation supports projects":** This and following quote from John Templeton Foundation, "Overview of Core Funding Areas," http://www.templeton.org/sites
/default/files/overview-cfa_0.pdf.
230 **One of the most horrible cases:** Details of King case from J. W. Brown 1988; Swan
and Swan 1988; Whiting 1989a, 1989b; Fraser 1999; Peters 2007; and Massachusetts
Citizens for Children, http://www.masskids.org/index.php?option=com_content
&id=161&Itemid=1652010.
232 **The children who have died:** Article on children of Jehovah's Witnesses who died
from refusing tranfusions at http://www.cftf.com/comments/kidsdied.html.
232 **In 1998, Seth Asser and Rita Swan:** Asser and Swan 1998; quotes that follow are
from that paper.
234 **It's not just the parents . . . Religious exemptions:** Lee, Rosenthal, and Scheffler
2013; CHILD 2011; National District Attorneys Association 2013; Massachusetts
Citizens for Children, DBRE [death by religious exemption], http://www.masskids
.org/index.php/federal-legislation-regarding-state-religious-exemption-laws.
235 **Islamic clerics in Afghanistan, Pakistan, and Nigeria:** Whitaker 2007; Katme 2011.
236 **"a respected figure":** A. M. Katme, "Islam, Vaccines, and Health," International Medical Council on Vaccination, http://www.vaccinationcouncil.org/2011/01/20/islam
-vaccines-and-health-2/.
236 **"the case of vaccination":** Brian Whitaker, "Is There a Doctor in the Mosque?," *Guardian,* http://www.theguardian.com/commentisfree/2007/May/11/doctorinthemosque.
236 **a study of childbirth in women:** Kaunitz et al. 1984.
236 **In 2013 . . . "We have no direct information":** Gerson 2013; CBC News, "Calgary
Strep Victim's Mother 'Will Be Held Accountable,' Say Police," November 23, 2013,
http://www.cbc.ca/news/canada/calgary/calgary-strep-victim-s-mother-will-be-held
-accountable-say-police-1.2437558.
237 **Within a given year:** Barnes et al. 2004; McCaffrey et al. 2004.
237 **The Vatican, for instance:** Bentson 2010; Keneally 2014.
238 **The work funded by NCCAM:** Mielczarek and Engler 2012; Offit 2012.
238 **While "alternative medicine" is often secular:** See Shermer 2013 for characteristics of pseudoscience.
239 **"Imagine a society's discovering a vaccine":** Stark 2001, p. 35 (emphasis in the original).
240 **It is because of this opposition:** History of stem cell restriction from Babington
2006; *Nature Cell Biology* 2010.
241 **One of the most egregious forms:** Information on HPV and vaccination at Centers
for Disease Control and Prevention, "HPV Vaccine—Questions & Answers," http://
www.cdc.gov/vaccines/vpd-vac/hpv/vac-faqs.htm, and at "Fact Sheet: Human Papillomavirus (HPV) Vaccines," http://www.cancer.gov/cancertopics/factsheet/preven
tion/HPV-vaccine.
241 **Surprisingly, in 2007 Governor Rick Perry:** Eggen 2011.
241 **Despite studies showing no apparent increase:** Bednarczyk 2012.
241 **"The seriousness of HPV and other STIs":** Focus on the Family, "Position Statement: HPV Vaccine," http://media.focusonthefamily.com/topicinfo/position_state
ment-human_papillomavirus_vaccine.pdf; see also Cole 2007.

242 **Although the Canadian government:** Craine 2012.

242 **Pressure from the Catholic Church:** Boodram 2013.

243 **But the situation isn't helped:** Ebola blamed on God and "homosexualism" from McCoy 2014; fasting, prayer, and President Sirleaf's statement from Front Page Africa, "Liberians Urged to Observe Three Days Fast and Prayer over Ebola," August 6, 2014, http://www.frontpageafricaonline.com/index.php/news/2589-liberians-urged-to-observe-three-days-fast-and-prayer-over-ebola.

244 **"Intentionally causing one's own death":** This and next quote from Declaration on Euthanasia, 1980, Sacred Congregation for the Doctrine of the Faith, http://www.vatican.va/roman_curia/congregations/cfaith/documents/rc_con_cfaith_doc_19800505_euthanasia_en.html.

244 **"I confirm that euthanasia is a grave violation":** John Paul II 1995.

245 **An AP-GfK poll:** AP-GfK poll, March 2014, http://ap-gfkpoll.com/main/wp-content/uploads/2014/04/AP-GfK-March-2014-Poll-Topline-Final_SCIENCE.pdf; see also Borenstein and Agiesta 2014.

246 **"When people are shown evidence":** Kahan 2014, p. 17.

247 **"We were put on this Earth as creatures of God":** Santorum quote reported in Hooper 2012.

247 **"'Never again will I curse the ground'":** John Shimkus on global warming in video at http://www.youtube.com/watch?feature=player_embedded&v=_7h08RDYA5E.

248 **Thirty-six percent of his fellow Americans:** Statistics from Public Relations Research Institute survey, "Americans More Likely to Attribute Increasingly Severe Weather to Climate Change, Not End Times," December 13, 2012, http://publicreligion.org/research/2012/12/prri-rns-december-2012-survey/.

248 **"What We Believe":** Cornwall Alliance for the Stewardship of Creation 2009.

249 **Since Americans with conservative attitudes:** Malka et al. 2012; Hirsh, Walberg, and Peterson 2013.

249 **profited financially from their efforts . . . "real scientific facts" . . . "real" evidence shows no trend:** Mervis 2011.

250 **"one of the greatest hoaxes":** Guest blogger 2009.

250 **"a massive international scientific fraud":** Piltz 2009.

250 **"to make people realize":** Consolmagno quoted in Harris Interactive 2013.

250 **Father Consolmagno's heart:** Iaccino 2014.

254 **The degree of pure atheism:** Religious belief in Europe from European Commission 2010; religious belief in America from Newport 2011.

254 **And on those scales:** Paul 2009; see also Zuckerman 2010 for an analysis of well-being in Scandinavian societies.

254 **My own analysis:** Coyne 2012.

255 **"I do not see in religion":** Napoleon 1916; the original quote, from *Au conseil d'état* (March 4, 1806), is "Je ne vois pas dans la religion le mystère de l'Incarnation mais le mystère de l'Ordre Social. La religion rattache au ciel une idée d'égalité qui empêche le riche d'être massacré par le pauvre."

255 **In the United States:** Norris and Inglehart 2004; Rees 2009; Delamontagne 2010; Solt, Habel, and Grant 2011.

256 **"*Religious* distress is at the same time":** Marx 2005, p. 175 (emphasis in the original).

256 **"Science without religion is lame":** Einstein 1954, p. 46.

257 **Einstein's views, often misconstrued:** Some examples of religion/science dialogues are those occurring regularly at the Institute on Religion in an Age of Science (http://www.iras.org), the International Society for Science and Religion (http://www.issr.org.uk/conferences/), and the Pontifical Academy of Sciences (http://www.casinapioiv.va/content/accademia/en/events.html).

257 **There are many theories:** See Boyer 2002 and Dennett 2006 for the variety of explanations.
259 **"The study of theology":** Paine 1848, part 2, p. 78.
260 **"I wish to propose":** Russell 1928, p. 1.
260 **"Pretending to be certain":** Harris 2005.
261 **The first involves:** Robert Park's story from Park 2008 (the quote is from p. 165) and my telephone interview with Park on October 14, 2014.
262 **Such was the case:** Briggs's story from van Biema 1998; Stauth 2013 (Briggs's quote on pp. 88–89); and "Taking Faith Healing Too Far," ABC *20/20*, January 6, 1999, transcript at http://www.skeptictank.org/hs/fhkiler1.htm.

REFERENCES

Alexander, D. D. 2010–11. "How Does a BioLogos Model Need to Address the Theological Issues Associated with an Adam Who Was Not the Sole Genetic Progenitor of Humankind?" biologos.org/uploads/projects/alexander_white_paper.pdf.

American Association for the Advancement of Science. 2006. "Statement on the Teaching of Evolution." http://archives.aaas.org/docs/resolutions.php?doc_id=443.

Angier, N. 2004. "My God Problem." *American Scholar* 73:131–34.

Armstrong, K. 2009. *The Case for God.* New York: Knopf.

Aslan, R. 2005. *No God but God: The Origins, Evolution, and Future of Islam.* New York: Random House.

Asser, S. M., and R. Swan. 1998. "Child Fatalities from Religion-Motivated Medical Neglect." *Pediatrics* 101:625–29.

Astin, J. A., et al. 2006. "The Efficacy of Distant Healing for Human Immunodeficiency Virus—Results of a Randomized Trial." *Alternative Theories in Health Medicine* 12:36–41.

Atkins, P. 1995. "Science as Truth." *History of the Human Sciences* 8:97–102.

Augustine. 2002. *Works of St. Augustine.* Vol. 13, *On Genesis.* Edited by J. E. Rotelle, translation and notes by E. Hill. Hyde Park, NY: New City Press.

Aus, M. 2012. "Conversion on Mount Improbable: How Evolution Challenges Christian Dogma." http://old.richarddawkins.net/articles/645853-conversion-on-mount-improbable-how-evolution-challenges-christian-dogma.

Aviles, J. M., et al. 2001. "Intercessory Prayer and Cardiovascular Disease Progression in a Coronary Care Unit Population: A Randomized Controlled Trial." *Mayo Clinic Proceedings* 76:1192–98.

Ayala, F. J. 2007. *Darwin's Gift to Science and Religion.* Washington, DC: Joseph Henry Press.

———. 2012. "Darwin and Intelligent Design." In *The Blackwell Companion to Science and*

Christianity, edited by J. B. Stump and A. G. Padgett, pp. 283–94. Oxford: Wiley-Blackwell.

Baars, B. J., W. P. Banks, and J. B. Newman, eds. 2003. *Essential Sources in the Scientific Study of Consciousness*. Cambridge, MA: Bradford Books/MIT Press.

Babington, C. 2006. "Stem Cell Bill Gets Bush's First Veto." *Washington Post*, July 20. http://www.washingtonpost.com/wp-dyn/content/article/2006/07/19/AR2006071900 524.html.

Baggini, J. 2011. "The Myth That Religion Is More About Practice Than Belief." *Guardian*, "Comment Is Free," December 9. http://www.guardian.co.uk/commentisfree/2011 /dec/09/myth-religion-practice-belief. Summary at http://julianbaggini.blogspot.com /2011/12/churchgoers-survey.html.

———. 2012a. "How Science Lost Its Soul and Religion Handed It Back." In *The Blackwell Companion to Science and Christianity*, edited by J. B. Stump and A. G. Padgett, pp. 510–19. Oxford: Wiley-Blackwell.

———. 2012b. "Atheists, Please Read My Heathen Manifesto." *Guardian*, "Comment Is Free," March 25. http://www.theguardian.com/commentisfree/2012/mar/25/atheists -please-read-heathen-manifesto.

Bains, S. 2011. "Questioning the Integrity of the John Templeton Foundation." *Evolutionary Psychology* 9:92–115.

Banerjee, K., and P. Bloom. 2013. "Would Tarzan Believe in God?: Conditions for the Emergence of Religious Belief." *Trends in Cognitive Sciences* 17:7–8.

Barbour, I. G. 2000. *When Science Meets Religion: Enemies, Strangers, or Partners?* San Francisco: HarperOne.

Barna Research. 2011. "Six Reasons Why Young Christians Leave the Church." Barna Group. https://www.barna.org/teens-next-gen-articles/528-six-reasons-young-christians-leave -church.

Barnes, P. M., et al. 2004. "Complementary and Alternative Medicine Use Among Adults: United States, 2002." *Advance Data*, no. 343:1–19.

Bartal, I. B.-A., J. Decety, and P. Mason. 2011. "Empathy and Pro-social Behavior in Rats." *Science* 334:1427–30.

Bartal, I. B.-A., et al. 2014. "Pro-social Behavior in Rats Is Modulated by Social Experience." *eLife* 3:10.7554/eLife.01385.

Bednarczyk, R. A., et al. 2012. "Sexual Activity–Related Outcomes After Human Papillomavirus Vaccination of 11- to 12-Year Olds." *Pediatrics* 130:798–805.

Benson, H., et al. 2006. "Study of the Therapeutic Effects of Intercessory Prayer (STEP) in Cardiac Bypass Patients: A Multicenter Randomized Trial of Uncertainty and Certainty of Receiving Intercessory Prayer." *American Heart Journal* 151:934–42.

Bentson, C. 2010. "Pope's Exorcist Says the Devil Is in the Vatican." ABC News, March 11. http://abcnews.go.com/Travel/chief-exorcist-rev-gabriele-amorth-devil-vatican/story? id=10073040.

Bernstein, A. A. 2006. "The Tragedy of Theology: How Religion Caused and Extended the Dark Ages." *Objective Standard* 1:11–37.

Bickel, B., and S. Jantz. 1996. *Bruce and Stan's Pocket Guide to Talking with God*. Eugene, OR: Harvest House.

Bishop, G. F., et al. 2010. "Americans' Scientific Knowledge and Beliefs About Human Evolution in the Year of Darwin." *Reports of the National Center for Science Education* 30:16–18.

Bloom, P. 2013. *Just Babies: The Origin of Good and Evil.* New York: Crown.

———. 2014. "Did God Make These Babies Moral? Intelligent Design's Oldest Attack on Evolution Is as Popular as Ever." *New Republic,* January 13. http://www.newrepublic .com/article/116200/moral-design-latest-form-intelligent-design-its-wrong.

Bonhoeffer, D. 1967. *Letters and Papers from Prison.* Edited by E. Bethge. New York: Macmillan.

Boodram, K. 2013. "'Denial of HPV Vaccines Will Cost More Lives.'" *Trinidad Express Newspapers,* February 1. http://www.trinidadexpress.com/news/_Denial_of_HPV _vaccines_will_cost_more_lives_-189463051.html.

Borenstein, S., and J. Agiesta. 2014. "Poll: Big Bang a Question for Most Americans." *The Big Story,* April 21. http://bigstory.ap.org/article/poll-big-bang-big-question-most-americans.

Boyer, P. 2002. *Religion Explained: The Evolutionary Origins of Religious Thought.* New York: Basic Books.

Branch, C. 1995. "The Jesus Seminar: The Slippery Slope on the Road to Heresy." *Watchman Fellowship* 12. http://www.watchman.org/articles/other-religious-topics/the -jesus-seminar-the-slippery-slope-to-heresy/.

Brosnan, S. F. 2011. "A Hypothesis of the Co-evolution of Cooperation and Responses to Inequity." *Frontiers in Neuroscience* 5:43.

———. 2013. "Justice- and Fairness-Related Behaviors in Nonhuman Primates." *Proceedings of the National Academy of Sciences* 110:10416–23.

Brown, D. 1991. *Human Universals.* New York: McGraw-Hill.

Brown, J. W. 1988. "'I'm in So Much Pain.' Transcripts Describe Young Christian Scientist's Agonizing Death." *Phoenix Gazette,* October 21, pp. A1, A4.

Carrier, R. 2010. "Christianity Was Not Responsible for Modern Science." In *The Christian Delusion,* edited by J. W. Loftus, pp. 396–419. Amherst, NY: Prometheus Books.

Carroll, S. 2012. "Does the Universe Need God?" In *The Blackwell Companion to Science and Christianity,* edited by J. B. Stump and A. G. Padgett, pp. 185–97. Oxford: Wiley-Blackwell.

———. 2013. "Science, Morality, Possible Worlds, Scientism, and Ways of Knowing." http:// www.preposterousuniverse.com/blog/2013/03/07/science-morality-possible-worlds-scien tism-and-ways-of-knowing/.

Charlton, B., and A. Verghese. 2010. "Caring for Ivan Ilyich." *Journal of General Internal Medicine* 25:93–95.

Chetty, R. 2013. "Yes, Economics Is a Science." *New York Times,* October 21, p. A19. http:// www.nytimes.com/2013/10/21/opinion/yes-economics-is-a-science.html?pagewanted =all&_r=0.

CHILD. 2011. "Religious Exemptions from Health Care for Children." http://childrens healthcare.org/?page_id=24#Exemptions.

Citizens for Objective Public Education. 2013. *COPE et al. v. Kansas State Board of Education et al.* Civil Action No. 13-4119-KHV-JPO. http://www.pacificjustice.org /uploads/1/3/1/7/13178056/complaint.filedstamped_-_cope.pdf.

Clarke, A. C. 1973. "Hazards of Prophecy: The Failure of Imagination." In *Profiles of the Future: An Enquiry into the Limits of the Possible.* New York: Harper & Row.

Cohen, N. 2014. "The Phantom Menace of Militant Atheism." *Guardian,* "Comment Is Free," September 6. http://www.theguardian.com/commentisfree/2014/sep/07/militant-atheism -religious-apologists-intellectuals.

Cole, E. 2007. "Conservatives Raise Red Flag on Mandatory HPV Vaccine for Girls." *Christian Post,* March 26. http://www.christianpost.com/news/conservatives-raise-red-flag-on-mandatory-hpv-vaccine-for-girls-26520/.

Collins, F. S. 2006. *The Language of God: A Scientist Presents Evidence for Belief.* New York: Free Press.

Conway Morris, S. 2003. *Life's Solution: Inevitable Humans in a Lonely Universe.* Cambridge, UK: Cambridge University Press.

Cornell University. 1892. "Charter of the Cornell University—Laws of New York, 1865, Chapter 585." In *Laws and Documents Relating to Cornell University, 1862–1892,* pp. 23–29. Ithaca, NY: Press of the Ithaca Democrat.

Cornwall Alliance for the Stewardship of Creation. 2009. "An Evangelical Declaration on Global Warming." http://www.cornwallalliance.org/2009/05/01/evangelical-declaration-on-global-warming/.

Cortical Rider. 2012. "Top 10 Reasons Science Is Another Religion." *Listverse,* December 15. http://listverse.com/2012/12/15/top-10-reasons-science-is-another-religion/.

Coyne, J. A. 2000. "Is NOMA a No Man's Land?" (Review of *Rocks of Ages: Science and Religion in the Fullness of Life,* by Stephen Jay Gould). *Times Literary Supplement,* June 9, pp. 28–29.

———. 2005. "The Faith That Dare Not Speak Its Name: The Case Against Intelligent Design." *New Republic,* August 22, pp. 21–33.

———. 2009a. *Why Evolution Is True.* New York: Viking.

———. 2009b. "Seeing and Believing." *New Republic,* February 4, pp. 32–41.

———. 2011. "Catholics Claim That Lies Are Truer Than Truth." *Why Evolution Is True* Web site, September 16. http://whyevolutionistrue.wordpress.com/2011/09/16/catholics-claim-that-lies-are-truer-than-truth/.

———. 2012. "Science, Religion, and Society: The Problem of Evolution in America." *Evolution* 66:2654–63.

———. 2013a. "You Can't Prove a Negative." *Why Evolution Is True* Web site, October 14. http://whyevolutionistrue.wordpress.com/2013/10/14/you-cant-prove-a-negative/.

———. 2013b. "Did Christianity (and Other Religions) Promote the Rise of Science?" *Why Evolution Is True* Web site, October 18. http://whyevolutionistrue.wordpress.com/2013/10/18/did-christianity-and-other-religions-promote-the-rise-of-science/.

———. 2013c. "No Faith in Science." *Slate,* November 14. http://www.slate.com/articles/health_and_science/science/2013/11/faith_in_science_and_religion_truth_authority_and_the_orderliness_of_nature.html?wpisrc=burger_bar.

Craig, W. L. 2012. "Sean Carroll on Science and God." http://www.reasonablefaith.org/sean-carroll-on-science-and-god.

Craine, P. 2012. "Despite Pressure Calgary Bishop Insists: No HPV Vaccine in Catholic Schools." *LifeSite News,* June 27. http://www.lifesitenews.com/news/calgary-bishop-under-fire-over-ban-of-hpv-vaccine-in-catholic-schools/.

Curry, O. S., D. A. Mullins, and H. Whitehouse. In preparation. "Are There Any Universal Moral Values?"

Dalai Lama. 2005. *The Universe in a Single Atom: The Convergence of Science and Spirituality.* New York: Morgan Road Books.

Darwin, C. 1859. *On the Origin of Species by Means of Natural Selection, or the Preservation of Favored Races in the Struggle for Life.* London: Murray.

———. 1871. *The Descent of Man, and Selection in Relation to Sex.* London: Murray.

Davies, P. 2007. "Taking Science on Faith." *New York Times,* November 24. http://www
.nytimes.com/2007/11/24/opinion/24davies.html?pagewanted=all&_r=0.

Dawkins, R. 1976. *The Selfish Gene.* Oxford: Oxford University Press.

———. 1997. "Is Science a Religion?" *Humanist,* January/February. http://www.thehumanist
.org/humanist/articles/dawkins.html.

———. 2006. *The God Delusion.* New York: Bantam.

Dehaene, S., and J.-P. Changeux. 2011. "Experimental and Theoretical Approaches to
Conscious Processing." *Neuron* 70:200–227.

Delamontagne, R. G. 2010. "High Religiosity and Societal Dysfunction in the United States
During the First Decade of the Twenty-First Century." *Evolutionary Psychology* 8:617–57.

Dembski, W. A. 2012. "Southern Baptist Journal: Is Darwinism Theologically Neutral?"
BioLogos. http://biologos.org/blog/southern-baptist-voices-is-darwinism-theologically
-neutral.

Dennett, D. 1991. *Consciousness Explained.* New York: Little, Brown.

———. 1995. *Darwin's Dangerous Idea.* New York: Simon & Schuster.

———. 2006. *Breaking the Spell: Religion as a Natural Phenomenon.* New York: Viking.

Dennett, D., and A. Plantinga. 2010. *Science and Religion: Are They Compatible?* New
York: Oxford University Press.

de Waal, F. 2006. *Our Inner Ape: A Leading Primatologist Explains Why We Are Who We
Are.* New York: Riverhead Trade.

———. 2013. *The Bonobo and the Atheist: In Search of Humanism Among the Primates.*
New York: Norton.

de Waal, F., S. Machedo, and J. Ober. 2009. *Primates and Philosophers: How Morality
Evolved.* Princeton, NJ: Princeton University Press.

Dittrich, L. 2013. "The Prophet." *Esquire,* August. http://www.esquire.com/features/the
-prophet.

Draper, J. W. 1875. *History of the Conflict Between Religion and Science.* London: Henry S.
King & Co.

Ecklund, E. H. 2010. *Science vs. Religion: What Scientists Really Think.* New York: Oxford
University Press.

Ecklund, E. H., and E. Long. 2011. "Scientists and Spirituality." *Sociology of Religion*
72:253–74.

Ecklund, E. H., and C. P. Scheitle. 2007. "Religion Among Academic Scientists: Distinctions, Disciplines, and Demographics." *Social Problems* 54:289–307.

Eggen, D. 2011. "Rick Perry Reverses Himself, Calls HPV Vaccine Mandate a 'Mistake.'" *Washington Post,* September 13. http://www.washingtonpost.com/politics/rick-perry-reverses
-himself-calls-hpv-vaccine-mandate-a-mistake/2011/08/16/gIQAM2azJJ_story.html.

Ehrman, B. 2013. *Did Jesus Exist?: The Historical Argument for Jesus of Nazareth.* New
York: HarperOne.

Einstein, A. 1954. "Science and Religion." In *Ideas and Opinions,* pp. 41–49. New York: Crown.

Enns, P. 2012. *The Evolution of Adam: What the Bible Does and Doesn't Say About Human
Origins.* Grand Rapids, MI: Brazos Press.

European Commission. 2010. *Special Eurobarometer Biotechnology Report.* http://ec
.europa.eu/public_opinion/archives/ebs/ebs_341_en.pdf.

Evans, E. 1956. "Tertullian's Treatise on the Incarnation." http://www.tertullian.org/articles/evans_carn/evans_carn_04eng.htm.

Feynman, R. 1986. *Report of the Presidential Commission on the Space Shuttle* Challenger Accident. "Volume 2: Appendix F—Personal Observations on Reliability of Shuttle." http://history.nasa.gov/rogersrep/v2appf.htm.

Feynman, R., and R. Leighton. 1985. "Cargo Cult Science." In *Surely You're Joking, Mr. Feynman*, pp. 338–46. New York: Norton.

Finch, C. E. 2010. "Evolution of the Human Lifespan and Diseases of Aging: Roles of Infection, Inflammation, and Nutrition." *Proceedings of the National Academy of Sciences of the USA* 107:1718–24.

Fishman, Y., and M. Boudry. 2013. "Does Science Presuppose Naturalism (or Anything at All)?" *Science and Education* 22:921–49.

Flam, F. 2014. "New NY Times Column Suggests Most Scientific Results Are Wrong." Knight Science Journalism Tracker, January 27. http://ksj.mit.edu/tracker/2014/01/new-ny-times-column-suggests-most-scient.

Forrest, B. 2000. "Methodological Naturalism and Philosophical Naturalism: Clarifying the Connection." *Philo* 3:7–29.

Francalacci, P., et al. 2013. "Low-Pass DNA Sequencing of 1200 Sardinians Reconstructs European Y-Chromosome Phylogeny." *Science* 341:565–69.

France, A. 1894. *Le Jardin d'Épicure*. Paris: Calmann-Levy.

Franciscan Clerics of Holy Name College. 1943. *The National Catholic Almanac*. Paterson, NJ: St. Anthony's Guild.

Fraser, C. 1999. *God's Perfect Child: Living and Dying in the Christian Science Church*. New York: Henry Holt.

Freeman, C. 2003. *The Closing of the Western Mind: The Rise of Faith and the Fall of Reason*. New York: Knopf.

Funk, R. W., and the Jesus Seminar. 1998. *The Acts of Jesus: The Search for the Authentic Deeds of Jesus*. San Francisco: Harper San Francisco.

Gallup. 2007. "Majority of Republicans Doubt Theory of Evolution." http://www.gallup.com/poll/27847/majority-republicans-doubt-theory-evolution.aspx.

———. 2011. "Christianity Remains Dominant Religion in the United States." http://www.gallup.com/poll/151760/christianity-remains-dominant-religion-united-states.aspx.

———. 2014. "Evolution, Creationism, Intelligent Design." http://www.gallup.com/poll/21814/evolution-creationism-intelligent-design.aspx.

Gallup G., Jr., and D. M. Lindsay. 1999. *Surveying the Religious Landscape*. Harrisburg, PA: Morehouse.

Galton, F. 1872. "Statistical Inquiries into the Efficacy of Prayer." *Fortnightly Review* 12:125–35.

Garrigan, D., and M. F. Hammer. 2006. "Reconstructing Human Origins in the Genomic Era." *Nature Reviews Genetics* 7:669–80.

Gerson, J. 2013. "Calgary Mother Who Relied on Herbal Medicines Facing Charges After Son, 7, Dies of Treatable Bacterial Infection." *National Post*, November 22. http://news.nationalpost.com/2013/11/22/calgary-mother-who-relied-on-herbal-medicines-facing-charges-after-son-7-dies-of-treatable-bacterial-infection/.

Giberson, K. W. 2009. *Saving Darwin: How to Be a Christian and Believe in Evolution*. New York: HarperOne.

Giberson, K. W., and F. S. Collins. 2011. *The Language of Science and Faith*. Downers Grove, IL: InterVarsity Press.

Gilkey, L. 1985. *Creationism on Trial: Evolution and God at Little Rock*. Minneapolis: Winston Press.

Gould, S. J. 1982. "Genesis vs. Geology." *Atlantic Monthly* 250:10–17.

———. 1983. "Evolution as Fact and Theory." In *Hen's Teeth and Horse's Toes: Further Reflections in Natural History*, pp. 253–62. New York: Norton.

———. 1987. "Darwinism Defined: The Difference Between Fact and Theory." *Discover*, January, pp. 64–70.

———. 1989. *Wonderful Life: The Burgess Shale and the Nature of History*. New York: Norton.

———. 1999. *Rocks of Ages: Science and Religion in the Fullness of Life*. New York: Ballantine.

Gould, S. J., and E. S. Vrba. 1982. "Exaptation—A Missing Term in the Science of Form." *Paleobiology* 8:4–15.

Grant, T. 2008. "Measuring Aggregate Religiosity in the United States, 1952–2005." *Sociological Spectrum* 28:470–76.

———. 2014. "The Great Decline: 61 Years of Religiosity in One Graph, 2013 Hits a New Low." Religion News Service. http://tobingrant.religionnews.com/2014/08/05/the-great-decline-61-years-of-religion-religiosity-in-one-graph-2013-hits-a-new-low/.

Grayling, A. C. 2013. *The God Argument: The Case Against Religion and for Humanism*. New York: Bloomsbury.

Gross, N., and S. Simmons. 2007. "How Religious Are America's College and University Professors?" *Social Science Research Council (SSRC)* 70:101–9.

———. 2009. "The Religiosity of American College and University Professors." *Sociology of Religion* 70:101–29.

Guest blogger. 2009. "Rep. Broun Receives Applause on House Floor for Calling Global Warming a 'Hoax.'" *ThinkProgress*. http://thinkprogress.org/politics/2009/06/26/47862/broun-globalwarming-hoax/.

Haack, S. 2012. "Six Signs of Scientism." *Logos and Epistime* 3, no. 1:75–95.

Haidt, J. 2001. "The Emotional Dog and Its Rational Tail: A Social Intuitionist Approach to Moral Judgment." *Psychological Review* 108:814–34.

———. 2013. *The Righteous Mind: Why Good People Are Divided by Politics and Religion*. New York: Vintage.

Hamelin, L. R. 2014. "The Limits of Science." *The Barefoot Bum*. http://barefootbum.blogspot.com/2014/02/the-limits-of-science.html.

Harris, S. 2004. *The End of Faith*. New York: Norton.

———. 2005. "An Atheist Manifesto." http://www.rationalresponders.com/an_atheist_manifesto_by_sam_harris.

———. 2006. *Letter to a Christian Nation*. New York: Knopf.

———. 2007. "Response to Jonathan Haidt's 'Moral Psychology and the Misunderstanding of Religion.'" *Edge*. http://www.edge.org/discourse/moral_religion.html#harris.

———. 2011. *The Moral Landscape: How Science Can Determine Human Values*. New York: Free Press.

Harris Interactive. 2013. "Americans' Belief in God, Miracles, and Heaven Declines." http://

www.harrisinteractive.com/NewsRoom/HarrisPolls/tabid/447/ctl/ReadCustom%20De
fault/mid/1508/ArticleId/1353/Default.aspx.

Hart, D. B. 2013. *The Experience of God: Being, Consciousness, Bliss.* New Haven, CT: Yale
University Press.

Haught, J. F. 2003. *Deeper Than Darwin: The Prospect for Religion in the Age of Evolution.*
Boulder, CO: Westview Press.

———. 2006. *Is Nature Enough?: Meaning and Truth in the Age of Science.* Cambridge, UK:
Cambridge University Press.

———. 2007. "The Atheist Delusion." *Salon,* December 18. http://www.salon.com/2007/12/19
/john_haught/.

Hauser, M. 2006. *Moral Minds: How Nature Designed Our Universal Sense of Right and
Wrong.* New York: Ecco.

Hazen, R. 2005. *Genesis: The Scientific Quest for Life's Origins.* Washington, DC: Joseph
Henry Press.

Henn, B. M., L. L. Cavalli-Sforza, and M. W. Feldman. 2012. "The Great Human Expansion." *Proceedings of the National Academy of Sciences* 109:17758–64.

Hess, W. R. 2009. *God and Evolution.* National Center for Science Education. http://ncse
.com/religion/god-evolution.

Hirsh, J. B., M. D. Walberg, and J. B. Peterson. 2013. "Spiritual Liberals and Religious Conservatives." *Social Psychology and Personality Science* 4:14–20.

Hitchens, C. 2003. "Mommie Dearest: The Pope Beatifies Mother Teresa, a Fanatic, a Fun-
damentalist, and a Fraud." *Slate,* October 20. http://www.slate.com/articles/news
_and_politics/fighting_words/2003/10/mommie_dearest.html.

———. 2007a. *God Is Not Great: How Religion Poisons Everything.* New York: Twelve.

———. 2007b. "An Atheist Responds." *Washington Post,* July 13. http://www.washington
post.com/wp-dyn/content/article/2007/07/13/AR2007071301461.html.

Hooper, T. 2012. "Santorum and Gingrich." *Colorado Independent,* February 6. http://www
.coloradoindependent.com/111924/santorum-and-gingrich-dismiss-climate-change-vow
-to-dismantle-the-epa.

Huff, T. E. 2003. *The Rise of Early Modern Science: Islam, China, and the West.* 2nd ed.
Cambridge, UK: Cambridge University Press.

Hume, D. 1975 [1748]. *Enquiries Concerning Human Understanding and Concerning the
Principles of Morals.* 3rd ed. Edited by L. A. Selby-Bigge and P. H. Niddich. Oxford:
Oxford University Press.

Humphrey, N. 1996. *Leaps of Faith: Science, Miracles, and the Search for Supernatural
Consolation.* New York: Basic Books.

Hutchinson, I. 2003. "Science: Christian and Natural." *Perspectives on Science and Christian Faith* 55:72–79.

———. 2011. *Monopolizing Knowledge: A Scientist Refutes Religion-Denying, Reason-
Destroying Scientism.* Belmont, MA: Fias Publishing.

Iaccino, L. 2014. "Pope's Astronomer Guy Consolmagno Says 'Aliens Exist.'" *International
Business Times,* October 7. http://www.ibtimes.co.uk/popes-astronomer-guy-consol
magno-says-aliens-exist-1467400.

Ingersoll, R. G. 1879. "The Gods." In *The Gods and Other Lectures.* Washington, DC:
C. P. Farrell.

———. 1900a. "God in the Constitution." In *The Works of Robert G. Ingersoll*. Dresden Edition, vol. 11, pp. 121–34. Washington, DC: C. P. Farrell.

———. 1900b. "The Gods." In *The Works of Robert G. Ingersoll*. Dresden Edition, vol. 1, pp. 7–90. Washington, DC: C. P. Farrell.

———. 1902. *The Works of Robert G. Ingersoll: Interviews*. Dresden Edition, vol. 8. Washington, DC: C. P. Farrell.

Ipsos/Reuters. 2011. "Global @dvisor: Supreme Being(s), the Afterlife, and Evolution." http://www.ipsos-na.com/download/pr.aspx?id=10670.

Isaacson, W. 2007. *Einstein: His Life and Universe*. New York: Simon & Schuster.

Jacoby, S. 2013. "An Interview with Susan Jacoby on Atheism." *Five Books*. http://five books.com/interviews/susan-jacoby-on-atheism.

James, W. 1928. *The Varieties of Religious Experience: A Study in Human Nature*. 36th impression. London: Longmans, Green & Co.

John Paul II. 1995. *Evangelium Vitae*. http://www.vatican.va/edocs/ENG0141/_INDEX .HTM.

———. 1996. "Message to the Pontifical Academy of Sciences: On Evolution." https://www .ewtn.com/library/PAPALDOC/JP961022.HTM.

Johnson, G. S. 2014. "New Truths That Only One Can See." *New York Times*, January 21. http:// www.nytimes.com/2014/01/21/science/new-truths-that-only-one-can-see.html?ref=today spaper&_r=1.

Jones, J. M. 2010. "Few Americans Oppose National Day of Prayer." Gallup. http://www .gallup.com/poll/127721/few-americans-oppose-national-day-prayer.aspx.

———. 2011. "In U.S., 3 in 10 Say They Take the Bible Literally." Gallup. http://www.gallup .com/poll/148427/say-bible-literally.aspx.

Kahan, D. M. 2014. "Climate Science Communication and the Measurement Problem." *Advances in Political Psychology*, in press.

Kass, L. 2007. "Keeping Life Human: Science, Religion, and the Soul." Wriston Lecture. http://www.manhattan-institute.org/html/wl2007.htm.

Katme, A. M. 2011. "Islam, Vaccines, and Health." International Medical Council on Vaccination. http://www.vaccinationcouncil.org/2011/01/20/islam-vaccines-and-health-2/.

Kaufmann, W. 1958. *Critique of Religion and Philosophy*. New York: Harper & Bros.

———. 1961. *The Faith of a Heretic*. New York: Doubleday & Co.

Kaunitz, A. M., et al. 1984. "Perinatal and Maternal Mortality in a Religious Group Avoiding Obstetric Care." *American Journal of Obstetrics and Gynecology* 150:826–31.

Keller, M. A., A. V. Turchyn, and M. Ralser. 2014. "Non-enzymatic Glycolysis and Pentose Phosphate Pathway-like Reactions in a Plausible Archean Ocean." *Molecular Systems Biology* 10:725.

Keneally, M. 2014. "Inside the Exorcist Group That's Won the Vatican's Blessing." ABC News, July 3. http://abcnews.go.com/International/exorcism-group-vaticans-bless ing/story?id=24412640.

Kieran, M. 2009. "Cognitive Value of Art." In *A Companion to Aesthetics*, 2nd ed., edited by S. Davies et al., pp. 194–97. Oxford: Blackwell.

Kitcher, P. 1982. *Abusing Science: The Case Against Creationism*. Cambridge, MA: MIT Press.

———. 1985. *Vaulting Ambition: Sociobiology and the Quest for Human Nature*. Cambridge, MA: MIT Press.

———. 2014. *Life After Faith: The Case for Secular Humanism.* New Haven, CT: Yale University Press.

Krucoff, M. W., et al. 2005. "Music, Imagery, Touch, and Prayer as Adjuncts to Interventional Cardiac Care: The Monitoring and Actualisation of Noetic Trainings (MANTRA) II Randomised Study." *Lancet* 366:211–17.

Lamarck, J. B. 1820. *Système analytique des connaissances positives de l'homme.* Paris: A. Belin.

Larson, E. J., and L. Witham. 1997. "Scientists Are Still Keeping the Faith." *Nature* 386:435–36.

Lee, E. O., L. Rosenthal, and G. Scheffler. 2013. "The Effect of Childhood Vaccine Exemptions on Disease Outbreaks." Center for American Progress. http://www.american progress.org/wp-content/uploads/2013/10/VaccinesBrief-1.pdf.

Lehrer, J. 2010. "The Truth Wears Off." *New Yorker,* December 13. http://www.newyorker .com/reporting/2010/12/13/101213fa_fact_lehrer.

Lewontin, R. C. 1997. "Billions and Billions of Demons" (Review of *The Demon-Haunted World: Science as a Candle in the Dark,* by Carl Sagan). *New York Review of Books* 44:28–32.

Li, H., and R. Durbin. 2011. "Inference of Human Population History from Individual Whole-Genome Sequences." *Nature* 475:493–97.

Linker, D. 2014. "Why Atheism Doesn't Have the Upper Hand over Religion." *The Week.* https://theweek.com/article/index/260172/why-atheism-doesnt-have-the-upper-hand -over-religion.

Livingstone, D. N. 2011. *Adam's Ancestors: Race, Religion, and the Politics of Human Origins (Medicine, Science, and Religion in Historical Context).* Baltimore: Johns Hopkins University Press.

Loftus, J. W. 2013. *The Outsider Test for Faith: How to Know Which Religion Is True.* Amherst, NY: Prometheus Books.

Luhrmann, T. M. 2012. *When God Talks Back: Understanding the American Evangelical Relationship with God.* New York: Knopf.

Luther, M. 1857. *The Table Talk of Martin Luther.* London: H. G. Bohn.

McCaffrey, A. M., et al. 2004. "Prayer for Health Concerns: Results of a National Survey on Prevalence and Patterns of Use." *Archives of Internal Medicine* 164:858–62.

McCoy, T. 2014. "'God Is Angry with Liberia,' Local Religious Leaders Say, Blaming Ebola on 'Homosexualism.'" *Washington Post,* August 6. http://www.washingtonpost.com /news/morning-mix/wp/2014/08/06/god-is-angry-with-liberia-ebola-is-a-plague/.

McKown, D. B. 1993. *The Mythmaker's Magic: Behind the Illusion of "Creation Science."* Amherst, NY: Prometheus Books.

Malka, A., et al. 2012. "The Association of Religiosity and Political Conservatism: The Role of Political Engagement." *Political Psychology* 33:275–99.

Manier, J. 2008. "The New Theology." *Chicago Tribune,* January 20. http://articles.chica gotribune.com/2008-01-20/features/0801120310_1_atheists-till-evolution.

Marx, K. 2005. "A Contribution to the Critique of Hegel's Philosophy of Right." In *Collected Works,* vol. 3, *Marx and Engels, 1843–1844.* Translated by Jack Cohen et al. New York: International Publishers.

Masci, D. 2007. "How the Public Resolves Conflicts Between Faith and Science." Pew Fo-

rum on Religion and Public Life. http://pewforum.org/science-and-bioethics/how
-the-public-resolves-conflicts-between-faith-and-science/.

———. 2009. "Public Opinion on Religion and Science in the United States." Pew Forum
on Religion and Public Life. http://pewforum.org/2009/11/05/public-opinion-on
-religion-and-science-in-the-united-states/.

Mencken, H. L. 1922. "The Scientist." In *Prejudices: Third Series*, pp. 269–70. Binghamton,
NY: Knopf.

Mendez, F. L., et al. 2013. "An African American Paternal Lineage Adds an Extremely
Ancient Root to the Human Y Chromosome Phylogenetic Tree." *American Journal of
Human Genetics* 92:454–59.

Mervis, J. 2011. "Ralph Hall Speaks Out on Climate Change." *Science Insider.* http://news
.sciencemag.org/2011/12/ralph-hall-speaks-out-climate-change.

Mielczarek, E. V., and B. D. Engler. 2012. "Measuring Mythology: Startling Concepts in
NCCAM Grants." *Skeptical Inquirer* 36:34–43.

Miller, K. R. 1999. *Finding Darwin's God: A Scientist's Search for Common Ground Between God and Evolution.* New York: Cliff Street Books.

———. 2008. *Only a Theory: Evolution and the Battle for America's Soul.* New York: Viking.

———. 2012. "Science, Religion Incompatible? Hot-Button Debate Features Dr. Kenneth
R. Miller, Dr. Michael Shermer." *Huffington Post,* March 7. http://www.huffingtonpost
.com/2012/03/07/science-religion-incompatible_n_1327263.html?ref=religion-science.

Mohler, A. 2011. "False Start? The Controversy over Adam and Eve Heats Up." *AlbertMo,*
August 22. http://www.albertmohler.com/2011/08/22/false-start-the-controversy-over
-adam-and-eve-heats-up/.

Mooney, C. 2010. "Spirituality Can Bridge Science-Religion Divide." *USA Today,* September 13. http://usatoday30.usatoday.com/news/opinion/forum/2010-09-13-column13
_ST_N.htm.

Moreland, J. P. 1989. *Christianity and the Nature of Science.* Grand Rapids, MI: Baker
Book House.

Murphy, T. W. 2002. "Lamanite Genesis, Genealogy, and Genetics." In *American Apocry-
pha: Essays on the Book of Mormon,* edited by D. Vogel and B. L. Metcalfe, pp. 47–77.
Salt Lake City: Signature Books.

Nanda, M. 2004. "Postmodernism, Hindu Nationalism, and 'Vedic Science'" (part 1). *Free-
thinkers.* http://www.mukto-mona.com/Articles/vedic_science_Mira.htm.

Napoleon. 1916. *Napoleon in His Own Words.* Edited by J. Bertaut, translated by H. E. Law
and C. L. Rhodes. Chicago: A. C. McClurg & Co.

National Academies. 2008. "Compatibility of Science and Religion." http://www.nas.edu
/evolution/Compatibility.html.

National District Attorneys Association. 2013. "Religious Exemptions to Child Neglect." http://
www.ndaa.org/pdf/Religious%20Exemptions%20to%20Child%20Neglect%202013.pdf.

National Science Teachers Association. 2013. "Position Statement: The Teaching of Evolution." http://www.nsta.org/about/positions/evolution.aspx.

Nature Cell Biology. 2010. "Editorial: Human Embryonic Stem Cell Research in the US:
Time for Change?" *Nature Cell Biology* 12:627.

Newport, F. 2011. "More Than 9 in 10 Americans Continue to Believe in God." Gallup.
http://www.gallup.com/poll/147887/Americans-Continue-Believe-God.aspx.

———. 2014a. "Mississippi and Alabama Most Protestant States in U.S." Gallup. http://www.gallup.com/poll/167120/mississippi-alabama-protestant-states.aspx.

———. 2014b. "In U.S., 42% Believe Creationist View of Human Origins." Gallup. http://www.gallup.com/poll/170822/believe-creationist-view-human-origins.aspx?ref=image.

Noordhof, P. 1995. "Scientism." In *The Oxford Companion to Philosophy*, ed. T. Honderich, p. 814. Oxford: Oxford University Press.

Norris, P., and R. Inglehart. 2004. *Sacred and Secular: Religion and Politics Worldwide.* Cambridge, UK: Cambridge University Press.

Nurbaki, H. T. M. B. 2007. *Verses from the Holy Qur'an and the Facts of Science.* New Delhi: Kitabbhvan.

Offit, P. A. 2012. "Studying Complementary and Alternative Therapies." *Journal of the American Medical Association* 307:1803–4.

Orr, H. A. 2005. "Why Intelligent Design Isn't." *New Yorker,* May 30, pp. 40–52.

Paine, T. 1848. *The Theological Works of Thomas Paine.* London: J. Watson.

Paley, W. 1809. *Natural Theology or Evidences of the Existence and Attributes of the Deity.* 12th ed. London: R. Faulder.

Park, R. L. 2008. *Superstition: Belief in the Age of Science.* Princeton, NJ: Princeton University Press.

Paul, G. 2009. "The Chronic Dependence of Popular Religiosity upon Dysfunctional Psychosociological Conditions." *Evolutionary Psychology* 7:398–441.

Paulson, S. 2010. *Atoms and Eden: Conversations on Religion and Science.* New York: Oxford University Press.

Pennock, R. T. 1999. *Tower of Babel: The Evidence Against the New Creationism.* Cambridge, MA: MIT Press.

Peters, S. F. 2007. *When Prayer Fails: Faith Healing, Children, and the Law.* Oxford: Oxford University Press.

Pew Research. 2009a. "Public Opinion on Religion and Science in the United States." Pew Research Religion and Public Life Project. http://www.pewforum.org/2009/11/05/public-opinion-on-religion-and-science-in-the-united-states/.

———. 2009b. "Scientists and Belief." Pew Research Religion and Public Life Project. http://www.pewforum.org/2009/11/05/scientists-and-belief/.

———. 2010. "U.S. Religious Knowledge Survey." Pew Forum on Religion and Public Life. http://www.pewforum.org/files/2010/09/religious-knowledge-full-report.pdf.

———. 2011. "Laws Penalizing Blasphemy, Apostasy and Defamation of Religion Are Widespread." Pew Research Religion and Public Life Project. http://www.pewforum.org/2012/11/21/laws-penalizing-blasphemy-apostasy-and-defamation-of-religion-are-widespread/.

———. 2012a. "'Nones' on the Rise." Pew Research Religion and Public Life Project. http://www.pewforum.org/2012/10/09/nones-on-the-rise/.

———. 2012b. "The World's Muslims: Unity and Diversity." Pew Research Religion and Public Life Project. http://www.pewforum.org/2012/08/09/the-worlds-muslims-unity-and-diversity-executive-summary/.

Philipse, H. 2012. *God in the Age of Science?: A Critique of Religious Reason.* Oxford: Oxford University Press.

Piltz, R. 2009. "Rep. Sensenbrenner Projects 'Fascism' and 'Fraud' onto Scientists, Is Rebutted at Hearing." Climate Science Watch. http://www.climatesciencewatch.org /2009/12/08/rep-sensenbrenner-projects-fascism-and-fraud-onto-scientists/.

Pinker, S. 2002. *The Blank Slate*. New York: Viking.

———. 2005. "So How *Does* the Mind Work?" *Mind and Language* 20:1–24.

———. 2007. "The Brain: The Mystery of Consciousness." *Time*, January 29. http://content .time.com/time/magazine/article/0,9171,1580394-1,00.html.

———. 2011. *The Better Angels of Our Nature: Why Violence Has Declined*. New York: Viking.

———. 2013. "Science Is Not Your Enemy: An Impassioned Plea to Neglected Novelists, Embattled Professors, and Tenure-less Historians." *New Republic*, August 19, pp. 28–33. http://www.newrepublic.com/article/114127/science-not-enemy-humanities.

Pius XII. 1950. "Encyclical *Humani Generis* of the Holy Father Pius XII." http:// www.vatican.va/holy_father/pius_xii/encyclicals/documents/hf_p-xii_enc_12081950_hu mani-generis_en.html.

Plantinga, A. 2000. *Warranted Christian Belief*. New York: Oxford University Press.

———. 2010. "Religion and Science." *Stanford Encyclopedia of Philosophy*. http://plato .stanford.edu/entries/religion-science/.

———. 2011. *Where the Conflict Really Lies: Science, Religion, and Naturalism*. Oxford: Oxford University Press.

Polkinghorne, J. 1994. *Science and Christian Belief: Theological Reflections of a Bottom-Up Thinker*. London: Society for Promoting Christian Knowledge.

———. 2011. *Science and Religion in Quest of Truth*. New Haven, CT: Yale University Press.

Polkinghorne, J., and N. Beale. 2009. *Questions of Truth: Fifty-One Responses to Questions About God, Science, and Belief*. Loiusville, KY: Westminster John Knox Press.

Pope, S. J. 2004. "The Evolutionary Roots of Morality in Theological Perspective." In *The Epic of Evolution: Science and Religion in Dialogue*, edited by J. B. Miller, pp. 189–98. Upper Saddle River, NJ: Pearson/Prentice Hall.

Popper, K. 1957. *The Open Society and Its Enemies*. 3rd ed. Vol. 2. London: Routledge & Kegan Paul.

Poznik, G. D., et al. 2013. "Sequencing Y Chromosomes Resolves Discrepancy in Time to Common Ancestor of Males Versus Females." *Science* 341:562–65.

Price, M. E. 2012. "Group Selection Theories Are Now More Sophisticated, but Are They More Predictive?" (Review of *A Cooperative Species: Human Reciprocity and Its Evolution*, by Samuel Bowles and Herbert Gintis). *Evolutionary Psychology* 10:45–49.

Pross, A. 2012. *What Is Life: How Chemistry Becomes Biology*. New York: Oxford University Press.

Ram, A. 2013. "Isro Chief Seeks Divine Help for Mars Mission." *Times of India*, November 5. http://timesofindia.indiatimes.com/india/Isro-chief-seeks-divine-help-for-Mars -mission/articleshow/25238936.cms.

Redlawsk, D. P. 2013. "Who I Am, and Who I Am Not." *New York Times*, August 8. http://www .nytimes.com/roomfordebate/2013/08/15/should-creationism-be-controversial/creation -and-evolution-beliefs-define-who-i-am-and-who-i-am-not.

Rees, T. J. 2009. "Is Personal Insecurity a Cause of Cross-National Differences in the Intensity of Religious Belief?" *Journal of Religion and Society* 11:1–24.

Rohde, D. 2003. "Her Legacy: Acceptance and Doubts of a Miracle." *New York Times*, October

20. http://www.nytimes.com/2003/10/20/world/her-legacy-acceptance-and-doubts-of-a -miracle.html.

Rosenberg, A. 2012. *The Atheist's Guide to Reality: Enjoying Life Without Illusions*. New York: Norton.

Rosenberg, N. A., et al. 2002. "Genetic Structure of Human Populations." *Science* 298:2381–85.

Russell, B. 1928. *Sceptical Essays*. London: George Allen & Unwin.

Sagan, C. 1996. *The Demon-Haunted World: Science as a Candle in the Dark*. New York: Random House.

———. 2006. *The Varieties of Scientific Experience: A Personal View of the Search for God*. New York: Penguin Press.

Sarewitz, D. 2012. "Sometimes Science Must Give Way to Religion." *Nature* 488:431.

Scaramanga, J. 2013. "Darwin Inspired Hitler: Lies They Teach in Texas." *Salon*, October 25. http://www.salon.com/2013/10/25/christian_textbooks_darwin_inspired_hitler/.

Schlitz, M., et al. 2012. "Distant Healing of Surgical Wounds: An Exploratory Study." *Explore* 8:223–30.

Scott, E. C. 1996. "Creationism, Ideology and Science." *Annals of the New York Academy of Science* 775:505–22.

Sheehan, S., K. Harris, and Y. S. Song. 2013. "Estimating Variable Effective Population Sizes from Multiple Genomes: A Sequentially Markov Conditional Sampling Distribution Approach." *Genetics* 194:647–62.

Shelley, P. B. 1915 [1813]. *The Necessity of Atheism*. In *Selected Prose Works of Shelley*. London: Watts & Co.

Shermer, M. 2002. *Why People Believe Weird Things: Pseudoscience, Superstition, and Other Confusions of Our Time*. New York: Henry Holt.

———. 2012. *The Believing Brain: From Ghosts and Gods to Politics and Conspiracies; How We Construct Beliefs and Reinforce Them as Truths*. New York: St. Martin's Griffin.

———. 2013. "Science and Pseudoscience: The Difference in Practice and the Difference It Makes." In *Philosophy of Pseudoscience: Reconsidering the Demarcation Problem*, edited by M. Pigliucci and M. Boudry, pp. 203–23. Chicago: University of Chicago Press.

Siegel, E. 2013. "How Many Planets Are in the Universe?" *Starts with a Bang*. http:// scienceblogs.com/startswithabang/2013/01/05/how-many-planets-are-in-the-universe/.

Singer, P. 1981. *The Expanding Circle: Ethics and Sociobiology*. New York: Farrar, Straus & Giroux.

Small, J. 2011. "The Common Ground Between Science and Religion." *Huffington Post*, August 30. http://www.huffingtonpost.com/jeffrey-small/the-battle-between-science -and-religion_b_938045.html.

Smith, T. 2012. "Beliefs About God Across Time and Countries." NORC/University of Chicago. www.norc.org/PDFs/Beliefs_about_God_Report.pdf.

Sniegowski, P. D. 1995. "The Origin of Adaptive Mutants: Random or Nonrandom?" *Journal of Molecular Evolution* 40:94–101.

Solt, F., P. Habel, and J. T. Grant. 2011. "Economic Inequality, Relative Power and Religiosity." *Social Science Quarterly* 92:447–65.

Spufford, F. 2012. "Dear Atheists . . ." *New Humanist* 127:34–36.

Stark, R. 2001. *One True God: Historical Consequences of Monotheism.* Princeton, NJ: Princeton University Press.

———. 2005. *The Victory of Reason: How Christianity Led to Freedom, Capitalism, and Western Success.* New York: Random House.

Stauth, C. 2013. *In the Name of God: The True Story of the Fight to Save Children from Faith-Healing Homicide.* New York: St. Martin's Press.

Stenger, V. 2007. *God: The Failed Hypothesis; How Science Shows That God Does Not Exist.* Amherst, NY: Prometheus Books.

Stenmark, M. 2012. "How to Relate Christian Faith and Science." In *The Blackwell Companion to Science and Christianity,* edited by J. B. Stump and A. G. Padgett, pp. 63–73. Oxford: Wiley-Blackwell.

Sullivan, A. 2011. "Must the Story of the Fall Be True?" *The Dish.* http://dish.andrewsullivan.com/2011/10/05/must-the-story-of-the-fall-be-true/.

Swan, R., and D. Swan. 1988. "Christian Science Parents Charged with Negligent Homicide in Arizona." Children's Healthcare Is a Legal Duty, Inc., newsletter, Summer.

Swinburne, R. 2004. *The Existence of God.* 2nd ed. Oxford: Oxford University Press.

———. 2012. "Arguments to God from the Observable Universe." In *The Blackwell Companion to Science and Christianity,* edited by J. B. Stump and A. G. Padgett, pp. 119–29. Oxford: Wiley-Blackwell.

Tavris, C., and E. Aronson. 2007. *Mistakes Were Made (But Not by Me).* Orlando: Harcourt.

Thompson, J. J. 1976. "Killing, Letting Die, and the Trolley Problem." *Monist* 59:204–17.

Trivers, R. 2011. *The Folly of Fools: The Logic of Deceit and Self-Deception in Human Life.* New York: Basic Books.

Tunggal, S. 2013. "Collaboration, Big Data, and the Search for the Higgs Boson." Transcript of talk by Andrzej Nowak, CERN openlab. http://www.slideshare.net/sumatunggal/collaboration-big-data-and-the-search-for-the-higgs-boson.

United States Catholic Conference. 1994. *Catechism of the Catholic Church.* http://www.vatican.va/archive/ENG0015/_INDEX.HTM.

van Biema, D. 1998. "Faith or Healing?" *Time,* August 31, pp. 68–69.

Voltaire and F.-M. Arouet. 1763. *The Works of M. de Voltaire. Translated from the French with Notes, Historical and Critical.* Vol. 26. London: Newbery, Baldwin, Johnson, Crowder Davies, Coote, Kearsley, and Collins.

Wallace, A. R. 1870. *The Limits of Natural Selection as Applied to Man. Contributions to the Theory of Natural Selection. A Series of Essays.* London and New York: Macmillan & Co.

Weaver, M. 2014. "Archbishop of Canterbury Admits Doubts About Existence of God." *Guardian,* September 18. http://www.theguardian.com/uk-news/2014/sep/18/archbishop-canterbury-doubt-god-existence-welby.

Weinberg, S. 1999. "A Designer Universe." *New York Review of Books* 46:46–48.

Whitaker, B. 2007. "Is There a Doctor in the Mosque?" *Guardian,* "Comment Is Free," May 11. http://www.theguardian.com/commentisfree/2007/may/11/doctorinthemosque.

White, A. D. 1932 [1896]. *A History of the Warfare of Science with Theology in Christendom.* 2 vols. New York: D. Appleton & Co.

White, J. 2012. "Jesus Wept: A Skeptic Faces Possible Charges for Debunking Mumbai's

Miracle Statue." *Slate,* July 7. http://www.slate.com/articles/health_and_science/new_scientist/2012/07/a_statue_of_jesus_oozing_holy_water_an_indian_skeptic_debunks_miracle.html.

Whitehead, A. N. 1925. *Science and the Modern World.* New York: Macmillan.

Whiting, B. 1989a. "Parents Draw Probation in Girl's Death." *Arizona Republic,* September 27, pp. A1, A6.

———. 1989b. "Plea Deal Lets Pair Avoid Felony." *Arizona Republic,* September 2, pp. B1, B5.

Wieland, C. 2000. "But the Bible's Not a Science Textbook, Is It?" Creation.com. http://creation.com/but-the-bibles-not-a-science-textbook-is-it.

Wieseltier, L. 2013. "Crimes Against Humanities: Now Science Wants to Invade the Liberal Arts. Don't Let It Happen." *New Republic,* September 3. http://www.newrepublic.com/article/114548/leon-wieseltier-responds-steven-pinkers-scientism.

Wigner, E. 1960. "The Unreasonable Effectiveness of Mathematics in the Natural Sciences." *Communications in Pure and Applied Mathematics* 13:1–14.

Wilson, E. O. 1975. *Sociobiology: The New Synthesis.* Cambridge, MA: Harvard University Press.

———. 1998. *Consilience: The Unity of Knowledge.* New York: Knopf.

Zuckerman, P. 2010. *Society Without God: What the Least Religious Nations Can Tell Us About Contentment.* New York: New York University Press.

INDEX

Faraday Institute for Science and Religion, 19, 129
federal headship model (Adam and Eve story), 129–30
federal science funding, 7, 18, 240
Feynman, Richard, 28–29, 38, 189, 221
Fibiger, Johannes, 30
fideism, 68
film, 190, 191
Finding Darwin's God (Miller), 102
fine-tuning argument, xix, 21, 156, 159, 160–66, 227
Fishman, Yonatan, 114
Focus on the Family, 241–42
Followers of Christ, 262, 263
Folly of Fools, The: The Logic of Deceit and Self-Deception in Human Life (Trivers), 179–80
Forrest, Barbara, 94–95
Foundational Questions in Evolutionary Biology initiative, 19–20
France, Anatole, 117
Francis, Pope, 68–69, 122
Frank, Neil, 249
Freeman, Charles, 214
free will, 15, 16, 130
Frum cult, 22
fundamentalism, xvii, 14, 16, 55, 88, 108, 110

Galileo, 2, 4, 5, 215
Galton, Francis, 115
Garden of Eden. *See* Adam and Eve's historicity
genetic engineering, 224
genetics
    genetic counterevidence for religious claims, 126–27, 131–32
    genetic drift, 139–40
    Lysenko and, 220–21
    See also DNA; mutations
Giberson, Karl, 46, 74, 119–20, 204–5
Gilkey, Langdon, 75
Gingerich, Philip, 102
global-warming denialism, xix–xx, 245–50, 251
GMOs, 224
God
    anthropomorphic perceptions of, 43, 47–49, 56, 63, 133
    as beyond description, 43–44, 49
    as interactive in the world, 22, 23
    personal connection to, 42

as supernatural agent, 41, 42–43
    See also deism; God's existence; theistic evolution
God Delusion, The (Dawkins), 21
God Is Not Great: How Religion Poisons Everything (Hitchens), 21
"god of the gaps" arguments, xiii, 153, 154–56, 161, 178, 226–27
    See also natural theology
God's existence, 44–49
    as answer to unsolved questions ("god of the gaps" arguments), xiii, 153, 154–56, 161, 178, 226–27
    and NOMA argument, 111
    possibility of scientific evidence for, 23–24, 118–19, 203–4
    See also natural theology
God: The Failed Hypothesis; How Science Shows That God Does Not Exist (Stenger), 21
golden rice, 224
Goldstein, Herbert, 101
Gould, Stephen Jay, 6, 182
    on compatibility of science and religion, 99
    on human inevitability, 144–45
    NOMA thesis, xviii, 4, 64–65, 106–12, 226
    scientific fact defined, 31
Gray, Asa, 81, 118
Grayling, Anthony, 110, 149
Greece, ancient, science in, 212, 213, 214
Greene, Brian, 20
Gyatso, Tenzin, Dalai Lama, 105–6

Haack, Susan, 196–97
Haidt, Jonathan, 62, 167–68
Hamelin, L. R., 23–24
Harris, Martin, 121
Harris, Sam, xii, xxi, 21, 190, 260
Hart, David Bentley, 49
Harvard University's Foundational Questions in Evolutionary Biology initiative, 19–20
Haught, John, 73–74, 79, 111, 135, 147–48, 206
health and disease. *See* disease and healing; medicine
hell, 54, 58, 71, 87, 88
    See also afterlife
Henry, Fred, 242
heresy, 58, 70, 72, 85, 214, 215
Higgs, Peter, 36
Higgs boson, 36, 39, 205, 209

King, Catherine, 230–31, 236
King, John, 230–31, 236
Kitcher, Philip, xxii, 86
knowledge
  defined, 186
  mathematical or philosophical,
    188–89, 198
  See also scientific knowledge; truth; ways
    of knowing
Kurosawa, Akira, *Ikiru*, 190, 191

Laestadianism, 84
Lamarck, Jean-Baptiste, 185
*Language of God, The: A Scientist Presents
  Evidence for Belief* (Collins), 185
Laplace, Pierre-Simon, 92, 149
law
  assisted dying, 243, 245
  religious privileging and exemptions, xviii,
    231, 234–36, 242, 263
  sharia law, 251–52
  stem cell research bans, 240
laws of nature. *See* physical laws
learning
  beliefs and, 181–82
  morality and, 170–71, 172, 176, 181
Letter to a Christian Nation (Harris), 21
Lewontin, Richard, 26, 92–93
liberal theology, xvii, 16, 44, 49, 75, 113, 129
  See also specific theologians
liberation theology, 87
Liberia, Ebola in, 243
life, origins of, 37, 146, 156, 157
*Life After Faith: The Case for Secular
  Humanism* (Kitcher), xxii
*Life's Solution: Inevitable Humans in a Lonely
  Universe* (Conway Morris), 142–43
linguistics, 39, 187, 199
Linker, Damon, 168, 175
literalism. *See* scriptural literalism
literature, 190–93
Livingstone, David, 56
Long, Elizabeth, 103
Lotfus, John, 85–86
Lovett, Ryan, 236
Lovett, Tamara, 236
Luhrmann, Tanya, 62
Luther, Martin, 69, 70, 72, 215
Lutheranism, 70
  Laestadianism, 84
lyrebirds, 184
Lysenko affair, 220–21

McKown, Delos, 80
marsupials, 142, 143, 149
Marx, Karl, 256
materialism, 91
mathematics, xii, 156, 158, 159, 184, 188, 198
Mécanique Céleste (Laplace), 92
medical neglect, xix, 229–39
  harm to adults, 233, 236
  harm to children, xix, 230–34, 236, 237,
    262–63
  and law, 231, 234–35, 237
medicine
  abnegation of, 5, 83, 149, 229, 242–43, 251
  assisted dying, 243–45
  science's advances and effectiveness, 5,
    206–7, 261–62
  stem cell research, xix–xx, 217, 240–41
  vaccinations, xix, 5, 217, 235–36, 241–42
  See also alternative medicine; disease and
    healing; medical neglect
Mencken, H. L., 38
metaphorical interpretations of scripture, 44,
  54–59, 74–75, 129–30
methodological naturalism, 92
  See also naturalism; science; scientific
    method(s)
methods, incompatibility between religious
  and scientific, 64, 65–89
  authority as arbiter of truth, 69–72
  cherry-picking, 74–77, 129–30
  fabricated answers to difficult/insoluble
    questions, 78–81
  progress vs. stasis, 87–89, 157
  religion's dependence on faith, 67–69
  See also counterevidence; evidence;
    falsifiability
Mill, John Stuart, 110
millennialism, 123
Miller, Kenneth, 100, 102, 137–38, 143, 160,
  202, 217
miracles, 109, 116–17, 120–24
missionizing, xvii, 229, 239
Mohler, Albert, 128
monkeys, 171
"Monkey Trial," 2, 5, 134
monogenism, 55–58, 105
  See also Adam and Eve's historicity
moral intuitions, 167–68, 169–70, 174–75
morality
  animal behavior and, 171–72, 175–76
  as argument for God's existence, 21, 156,
    168–77, 226

conflicting moral codes in different
religions, 84
cultural differences, 170, 176, 189
as key feature of religion, 42, 219–20
near-universality of, 169–70, 189
and NOMA argument, xviii, 107, 110–11
nonbelief and, 91, 189–90, 250, 252, 255, 261
scientific study of/explanations for, 15,
134, 166–68, 171–76, 189–90, 199–200
scientism and, 197, 198
secular ethics discourse, 110, 111, 189
socially driven changes, 88, 170–71, 177, 189
as subjective, 189–90
See also evil and suffering; religious harm;
scientific ethics
Moreland, J. P., 89
Mormonism, 62, 71, 82, 83, 125, 131–32
Moroni, 82, 125
Muhammad, 44, 53, 66, 83, 259
multiverse theory, 163
Mumbai Jesus statue, 122
music. *See* arts and humanities
mutations, 106, 133, 137, 138, 146–47
suffering caused by, 81, 148
See also natural selection

Nanda, Meera, 105
Napoleon Bonaparte, 92, 255
National Academies, 112
National Academy of Sciences, 12, 249–50
National Center for Complementary and
Alternative Medicine, 238
National Center for Science Education, 8, 9,
93–94
National Science Foundation, 18
National Science Teachers Association, 112
Native American origins, 82, 125, 131–32
natural disasters, 248
naturalism
methodological, 91–95
philosophical, 37, 94–95
seen as based on faith, 204–11
See also science; scientific method(s)
naturalistic fallacy, 111
natural laws. *See* physical laws
natural selection, 78, 92, 106, 133–34,
138–39, 144–45
altruism and, 174–75
human intelligence and, 134, 183–86
suffering inherent in, 81, 147, 148
See also evolutionary biology; evolution
entries

natural theology, 21, 105, 152–85
before Darwin, 77–78, 153–54, 213
consciousness argument, 158
evaluating natural theological
explanations, 156–57
fine-tuning argument, xix, 156, 159,
160–66, 227
"god of the gaps" arguments, xiii, 153,
154–56, 161, 178, 226–27
human inevitability argument, 140–47,
156, 157, 226
"Moral Law" argument, 156,
168–77, 226
origin-of-life argument, 156, 157
overviews, xix, 151, 153–56
physical and mathematical laws
arguments, 156, 158–60, 227
seen as progenitor of science, 211–12
superfluous intelligence argument, 134,
183–86
true beliefs/rationality argument, 156,
177–83, 226
"you can't prove God doesn't exist"
argument, 152, 201–4
Natural Theology (Paley), 154
Nature, 205
NCCAM (National Center for
Complementary and Alternative
Medicine), 238
neuroscience, 15, 158, 199
neutrinos, 35
New Atheism, xii–xiii, 14, 20–21
See also atheism
new natural theology. *See* natural theology
Newton, Isaac, 92, 99, 153, 159, 213, 216
Nicene Creed, 49–51, 70–71
Noah's flood, 1, 57–58, 87, 104, 124
Nobel Prize, 30
NOMA (non-overlapping magisteria)
argument, xviii, 4, 64–65, 106–12,
196, 226
nonbelievers
as accommodationists, 98
nontheistic modern Europe, 250,
254–55, 261
scientists as, xviii, 12–14, 95, 216
statistics, 9, 12–13, 16, 47
See also agnosticism; atheism; nonreligious
societies; nontheistic religion; religiosity
non-overlapping magisteria (NOMA)
argument, xviii, 4, 64–65, 106–12,
196, 226

nonreligious societies
  considering a world without faith,
    xxi–xxii, 250–56, 260–61
  modern northern Europe, 252,
    254–55, 261
  Stalinist Soviet Union, 220–21
nontheistic religion and spirituality, xvi, 42,
  65, 111
  deism, 42, 99, 135
  pantheism, 65, 100, 101–2
Nowak, Martin, 19
N rays, 30
nuclear weapons, 111, 152, 217, 218
Nurbaki, Halûk, 105

Obama, Barack, 240
observation, 32–33, 65, 153
  See also scientific method(s)
On the Origin of Species (Darwin), 2, 14, 33,
  70, 154
original sin, 21–22, 54, 56, 124, 125–26,
  128–29, 130
  See also Adam and Eve's historicity;
  salvation
other ways of knowing. *See* ways of knowing
"Outsider Test for Faith" (OTF), 85–86

Paine, Thomas, 259
Paley, William, 154
pantheism, 65, 100, 101–2
paradise, 57, 84
  See also afterlife
paranormal phenomena. *See* supernatural
  and paranormal phenomena
parasites, 139
Park, Robert L., 261–62
parrots, 184
parsimony, 36–37, 122
Paul, the Apostle, Saint, 68, 121, 126,
  130–31, 260
Pennock, Robert, 113
Pentecostalism, 22
Perry, Rick, 241
Philipse, Herman, 123, 153, 165
philosophical naturalism, 37, 94–95
philosophical religions. *See* nontheistic
  religion and spirituality
philosophy, secular, xii, 110, 111, 154, 168–69
  as way of knowing, 65, 188–89, 198
  See also specific philosophers
physical laws
  and anthropic principle, 159, 160–66

faith seen as required for acceptance of,
  204–5, 210
seen as arising from Christianity, 212–13
seen as God's creation, 135, 156,
  158–60, 226
physics, 16, 145–46, 150, 156
  multiverse theory, 163
  quantum mechanics, 137–38, 145–47, 159
Piltdown Man, 223
Pinker, Steven, 53, 169, 170, 172, 179, 199
Pius XII, Pope, 55–56, 71, 128–29, 133
Plantinga, Alvin
  on Bible's literal truth, 62–63
  on complex intelligence, 184
  on naturalistic evolution, 135
  on nature of God, 48–49, 63
  sensus divinitatis notion, 177–79, 180–81
  on suffering, 148–49
Plato, 189
plumbing, 40–41
polio, 235
politics
  global-warming denialism and, 246–47
  religiosity and, 7, 249
Polkinghorne, John, 46, 79, 88–89
polygenism
  religious rejection of, 55–58, 105
  scientific proof of, 126–27
  See also human evolution
Pope, Stephen J., 103
Popper, Karl, 221
Porco, Carolyn, 208
postmodernism, 204, 222
prayer, 114, 115–16, 237–39
  See also faith-based healing
progress
  directionality in evolution, 135–36, 138–40
  scientific, vs. religious stasis, 86–89, 157
proselytizing, xvii, 229, 239, 253
psychology, 168–69, 187, 188
  See also evolutionary psychology

quantitativeness and quantitative methods,
  33, 36, 196
  See also mathematics
quantum mechanics, 137–38, 145–47, 159
Quran, 83, 105, 134, 259
Quranic literalism, 53, 58, 75, 76, 106

rationality. *See* reason and rationality
*Rationality in Science, Religion, and
  Everyday Life* (Stenmark), 45

revelation
as "another way of knowing," 150, 186, 195
the arts and, 193
faith and, 25, 61, 65, 67, 153, 210
Rocks of Ages: Science and Religion in the Fullness of Life (Gould), 106, 107
Rome, ancient, science in, 212, 213, 214
Roosevelt, Franklin, 218
Rosenberg, Alex, 198
Russell, Bertrand, xii, 260

Sagan, Carl, 93, 118, 119, 203
salvation, 15–16, 84, 125–26
See also original sin
Santorum, Rick, 247
Sarewitz, Daniel, 205, 209
Sartre, Jean-Paul, xii
Satan, 149
Sceptical Essays (Russell), 260
science, 27–41
and authority, 65, 70, 72, 205, 208–9
confirmation bias in, 29
denigration of, 151–52
harm done by accommodationism, 226–28
as method(s), 27, 28–29, 187
misuses of/harm caused by, xix, 107, 110–11, 152, 217–21
morality and, 107, 110–11, 212, 216–19
naturalism as guiding principle, 91–95, 216
natural theology as, 154
in NOMA view, xviii, 64, 106, 107, 110–11
pre-Christian and non-Christian, 212, 214
as pursuit of truth, xii, xx, 1, 5, 28–29, 187
rise of modern science in Europe, 212–16
scientific study of religion, 257–58
supernatural explanations in, 92, 93–94
as way of knowing, 93–94, 185, 187, 195, 198, 206–7, 222–24
See also scientific ethics; scientific knowledge; scientific method(s); scientism
science, criticisms of
"religion gave rise to science" claim, xix, 151, 211–17
"science can't prove that God doesn't exist" claim, 152, 201–4
"science does bad things" claim, 152, 217–21
"science is based on faith" claim, xix, 69, 152, 204–11
"science is fallible and unreliable" claim, 151–52, 222–24

science organizations, as promoters of accommodationism, 7–8, 19, 93–94, 98, 112
See also specific organizations
science/religion incompatibility. *See* religion/science incompatibility
science research funding, 7, 18, 19–20, 228, 240
scientific creationism, 14, 103–4
scientific ethics
science as morally neutral, 218–19
seen as arising from Christian morality, 212, 216–17
unethical misuses of science, xix, 107, 110–11, 152, 217–21
scientific knowledge, 187–89, 198
ability to perceive, seen as gift from God, 178
nature of scientific proof, 30–31, 32–33
as progressive and cumulative, 86–89, 157
as provisional, 28, 30–31, 95, 113–14, 117, 119, 151–52, 202, 222–24
seen as fallible/unreliable, 151–52, 222–24
See also evidence; scientific method(s)
scientific method(s), 31–41, 86, 187, 198
collectivity, 38–39
criticality and doubt in, 26–27, 34–35, 38, 65, 94
falsifiability, 33–34, 65
hypothesis formation and testing, 31–33
parsimony, 36–37
replication and quality control, 35–36
toleration of uncertainty, 37–38
uses outside hard science, 39–41, 187–88
See also evidence; falsifiability; methods, incompatibility between religious and scientific
scientism, xix, 53–54, 152, 185, 196–224
scientist-believers, xviii, 14, 95
as accommodationists, 98, 226
as argument for faith as basis of science, 213, 216
nontheist spiritual views, 101–3
prominent individuals, 99–100
statistics, 12
See also specific individuals
scientists
as arrogant, 198, 227
as nonbelievers, xviii, 12–14, 95, 216
as promoters of accommodationism, 6–9, 98

Swan, Rita, 232
Swinburne, Richard, 45, 48
syncretism, 100–106

Taoism, 42, 65
tapeworms, 139
Tasmanian devil, 149
Teilhard de Chardin, Pierre, 223
Templeton, John, 17–18
Templeton Foundation, 8, 14, 17–20, 103, 228
Templeton Prize, 18–19, 45, 46
Tennyson, Alfred, Lord, 1
Teresa, Mother, 117
terminal illness
    faith and, 252–53
    opposition to assisted dying, 243–45
Tertullian, 68
Thacker, Justin, 73
theism. *See* faith; religion *entries*
theistic evolution, 132–40, 226–27
    acceptance rates, xiii, 60, 132
    as Catholic Church's position, 61, 132, 133
    human evolution seen as inevitable, 140–47, 156
    logical problems with, 150
    range of views on, 135–38
    scientific problems with, 138–40, 143–44, 145–47
    theological problems with, 147–49
    See also human inevitability
theodicy, 81, 148–49
theology, xvii, 209, 259
    apologetics and rationalization in, 66–67, 75, 77–81, 88–89, 153, 157
    literalism in Christian theology, 57–59
    stasis and change in, 87–89
    See also natural theology; *specific theologians*
theory of everything, 159
Thomas Aquinas, Saint, 55, 57–58, 215
Thompson, Judith Jarvis, 169–70
Thomson, J. J., 99
Tolstoy, Leo, *The Death of Ivan Ilyich*, 191–92
transubstantiation, 61, 84
Trivers, Robert, 179
trolley problem, 169–70
true beliefs/rationality argument for God's existence, 156, 177–83, 226
Truman, Harry, 218
truth
    defined, 29–30, 186
    religion and, xii, xx–xxi, 43–46

science as pursuit of, xii, xx, 1, 5, 28–29, 187
    See also knowledge; religious claims; scientific knowledge; ways of knowing
Tyson, Neil deGrasse, xi

uncertainty, 37–38
    See also certainty; doubt
universe
    end of, 164
    multiverse theory, 163
    origins of, 28, 32, 37, 146, 245
Universe in a Single Atom, The (Dalai Lama), 105–6
universities
    church condemnation of the University of Paris (1277), 215
    science and religion institutes at, 6, 19
    university secularism, 3–4
U.S. Congress, 7
    House Committee on Science, Space, and Technology, 250
U.S. National Academy of Sciences, 12, 249–50

vaccinations, xix, 5, 217, 235–36, 241–42
Vander Woude, Thomas, 168, 174–75
Varieties of Religious Experience, The (James), 42, 44, 118
Varieties of Scientific Experience, The (Sagan), 118
Voltaire, 29
Vrba, Elisabeth, 182

Waal, Frans de, 171
Wallace, Alfred Russel, 134, 183
war, 111, 152, 217–18, 221
Watson, James D., 216
ways of knowing, 185–96
    the arts and humanities as, 185–86, 190–94
    hard science as, 93–94, 185, 187, 195, 198, 206–7, 222–24
    knowledge and truth defined, 186–87
    mathematics and philosophy as, 188–89, 198
    morality as, 189–90
    NOMA (non-overlapping magisteria) view, xviii, 4, 64–65, 106–12, 196
    "other ways of knowing" trope, xvi, xix, 24–25, 66, 150–52, 185, 186, 195–96, 227–28
    social sciences as, 186, 187–88, 195, 198